P9-CCO-540

WORKOUT FOR A
BALANCED
BRAIN

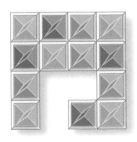

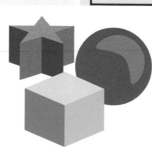

WORKOUT FOR A
BALANCED
BRAIN

Philip Carter and Ken Russell

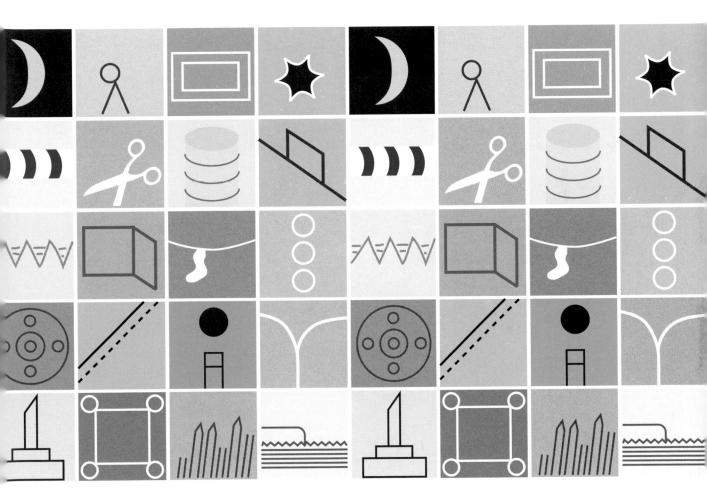

FOREWORD BY DR. CHRISTOPHER MARTYN

 Reader's
Digest

THE READER'S DIGEST ASSOCIATION, INC.
Pleasantville, New York / Montreal

A READER'S DIGEST BOOK

Conceived, designed, and produced by
Quarto Publishing plc
The Old Brewery
6 Blundell Street
London N7 9BH

Senior Project Editor Nicolette Linton
Senior Art Editor Elizabeth Healey
Text Editors Ian Kearey, Deirdre Clark
Designer Rod Teasdale
Illustrators Jenny Dooge, Sophie Joyce
Indexer Pamela Ellis

Art Director Moira Clinch
Publisher Piers Spence

First published in 2001 by
The Reader's Digest Association, Inc.
Pleasantville, NY 10570-7000

Copyright © 2001 Quarto Inc.

All rights reserved. Unauthorized reproduction, in any manner, is prohibited.

Reader's Digest and the Pegasus logo are registered trademarks of
The Reader's Digest Association, Inc.

READER'S DIGEST PROJECT STAFF
Project Editor Kimberly Ruderman
Associate Editorial Director Marianne Wait
Editorial Manager Christine R. Guido

Contributing Project Designer Jane Wilson

READER'S DIGEST ILLUSTRATED REFERENCE BOOKS
Editor-in-Chief Christopher Cavanaugh
Art Director Joan Mazzeo
Director, Trade Publishing Christopher T. Reggio
Senior Design Director, Trade Elizabeth L. Tunnicliffe
Editorial Director, Trade Susan Randol

Library of Congress Cataloging in Publication Data
Carter, Philip J.
  Workout for a balanced brain: exercises, puzzles & games to sharpen
both sides of your brain / Philip Carter & Ken Russell.
    p. cm.
  Includes bibliographical references and index.
  ISBN 0-7621-0331-0
    1. Lateral thinking puzzles.
2. Left and right (Psychology) I. Russell, Kenneth A. II. Title.

LB1507.L37 C37 2001
793.7—dc21   2001019610

Manufactured by Regent Publishing Services Ltd, Hong Kong
Printed by Star Standard, Singapore

13 5 7 9 10 8 6 4 2
QUAR.WBB

Contents

Foreword

→**DR. ROGER SPERRY**
His split-brain research forms the basis of our understanding of brain-hemisphere functions.

In 1981 American psychologist Roger Sperry was awarded a Nobel Prize in medicine for his discoveries concerning the functional specialization of the cerebral hemispheres. He and his colleagues had carried out a remarkable series of experiments on patients who had undergone brain surgery for epilepsy that was difficult to treat.

To stop the seizures from spreading, neurosurgeons cut the corpus callosum—the bridge of millions of nerve fibers that connects the two hemispheres of the brain and allows them to communicate with each other. After the operation, these patients lacked the ability to transfer information from one side of the brain to the other. Only the hemisphere that directly received a stimulus could process it.

When a picture was shown to one of these patients in such a way that the information reached only the left hemisphere, she could describe what she saw. But when the same image was presented to the right hemisphere, the patient denied seeing anything. Astonishingly though, when asked to point out an object similar to the one in the projected image, she could do so without difficulty.

It was clear that the right side of the brain had seen the picture and had recognized it. What it couldn't do was talk about what it had seen. As the experiments continued, they revealed much more about how the two sides of the brain were specialized to perform different intellectual tasks. The left hemisphere excelled at language, speech, and logic; the right hemisphere at pattern recognition, music, and emotion.

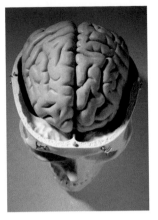

← **THE HUMAN BRAIN**
The control center for all bodily activities and the organ of thought.

Of course, most people have a corpus callosum that is intact and functioning. But recent advances in the techniques available for imaging the living nervous system have shown that much of what was deduced from the experiments on patients with split brains also holds for people whose brains are normal.

There is more than a grain of truth in the idea that writers use the left side of their brains, while graphic artists use the right. Just as there is some truth in the view that education systems in the western world tend to develop the analytic capacities of the left hemisphere at the expense of the creative talents of the right.

In this book, Philip Carter and Ken Russell have compiled an intriguing collection of tests and exercises. If you work through them, you'll discover more about the way you think. You'll find out whether you're left-brained or right-brained and you'll be able to identify the type of problem that you're good—or not so good—at solving. It may reveal unknown talents and allow you to develop abilities that you didn't know you possessed.

Reading this book may not turn you into a genius, but it will certainly provide an insight into the mental skills that are your strength and give you a chance to work on those that don't come naturally. At the very least, the exercises are fun and interesting to do.

Dr. Christopher Martyn M.A., D. Phil., FRCP
Senior Clinical Scientist, Medical Research Council
Environmental Epidemiology Unit,
Southampton University, England

Consulting Neurologist, Wessex Neurological Center
Southampton General Hospital, England

Hemispheres of the Brain

TWO SEPARATE SIDES OF THE BRAIN

"Each hemisphere of the human brain has its own private sensations, perceptions, thoughts and ideas, all of which are cut off from the corresponding experiences in the opposite hemisphere.... In many respects, each disconnected hemisphere appears to have a separate mind of its own."

Dr. Roger Sperry, Nobel Prize in medicine 1981

Science tells us that we only use 3 percent of our brain. This is the amount of information available to us consciously, and the rest is locked within the subconscious mind. Most of the time, we only utilize what is in the conscious left hemisphere of the brain, and this enables us to tackle the daily stresses of life without access to the vast amount of creativity, memory, and intuition found within the subconscious right hemisphere.

It has always been accepted that we human beings are all different in our own way—in other words, we are all individuals. We each have our own physical makeup, fingerprints, DNA, facial features, personality, dreams, and aspirations. Throughout history, these characteristics have been analyzed and categorized, but it was only recently realized that each one of us has two sides to our brain, each of which has quite different functions and attributes.

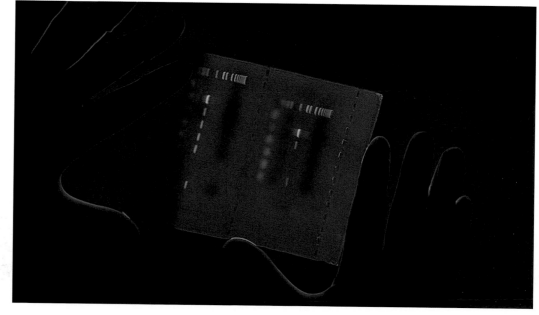

→**THE SKELETON OF THE DNA** consists of two chains of alternate sugar and phosphate groups.

AN ANCIENT CONCEPT?

The Chinese had touched on this thousands of years ago. They called it yin and yang, the two basic contrary forces in ancient Chinese thought, elaborated in Han Dynasty Confucianism. Yang is associated with the male side, characterized by cold, light, heaven, creation, dominance; and yin is the female side associated with warmth, darkness, earth, sustenance, and passivity. These two forces were alleged to exist in most things and operate cyclically to produce change.

Within the yin and yang symbol you can see a light dot in the dark side and a dark dot in the light side, indicating that in each male there is a part of him that exhibits female characteristics, and vice versa. The ancient Chinese did not hit completely on the notion of laterality, but they did get remarkably close to the correct analysis of what makes each one of us tick.

In reality, what we do now know is that the cerebral cortex has two halves. These are separate hemispheres that are connected by nerve fibers, the corpus callosum. The left side of the brain connects to the right side of the body, while the right brain connects to the left side—and each hemisphere controls different functions.

← **WINTER** is the season that represents yin—the feminine side—yet yin also includes warmth of heart.

→ **MOST TOP ATHLETES** possess a degree of aggression that is essential to be a winner.

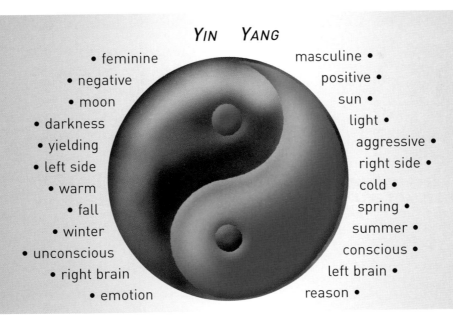

YIN   YANG

Yin	Yang
• feminine	masculine •
• negative	positive •
• moon	sun •
• darkness	light •
• yielding	aggressive •
• left side	right side •
• warm	cold •
• fall	spring •
• winter	summer •
• unconscious	conscious •
• right brain	left brain •
• emotion	reason •

Hemispheres of the Brain

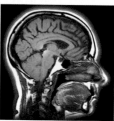

The term laterality—or sidedness—is used to refer to any one of a number of preferences for one side of the body to another. Probably the most common example of this, and one to which we can all relate, is whether a person is left- or right-handed. The brain also exhibits laterality; we now know that the two halves of the brain differ significantly with regard to specific cognitive—or thinking—functions, and that each of us has a distinct preference for one side or the other.

If we were to remove a brain from the skull, we would see two, almost identical, hemispheres. Each of these two halves has developed specialized functions and has its own private sensations, perceptions, ideas, and thoughts, all separate from the corresponding experiences in the opposite hemisphere.

↑ **NEWTON BY WILLIAM BLAKE**
As an author and artist, Blake exhibits a balance of the rational and intuitive sides of the brain.

SEQUENTIAL VERSUS HOLISTIC

In most of us, language and speech functions are mediated in the left hemisphere, and it is these linguistic functions that are most clearly associated with laterality. However, there is now a growing body of evidence to suggest that other cognitive and perceptual components of behavior are lateralized, in particular that the left hemisphere is analytical and functions in a sequential and rational fashion, whereas the right hemisphere is synthetic and functions in a holistic manner. The thought processes of the left hemisphere are characterized by order, sequence, and logic. In contrast, the right hemisphere controls spatial ability, artistic appreciation, and creative thought.

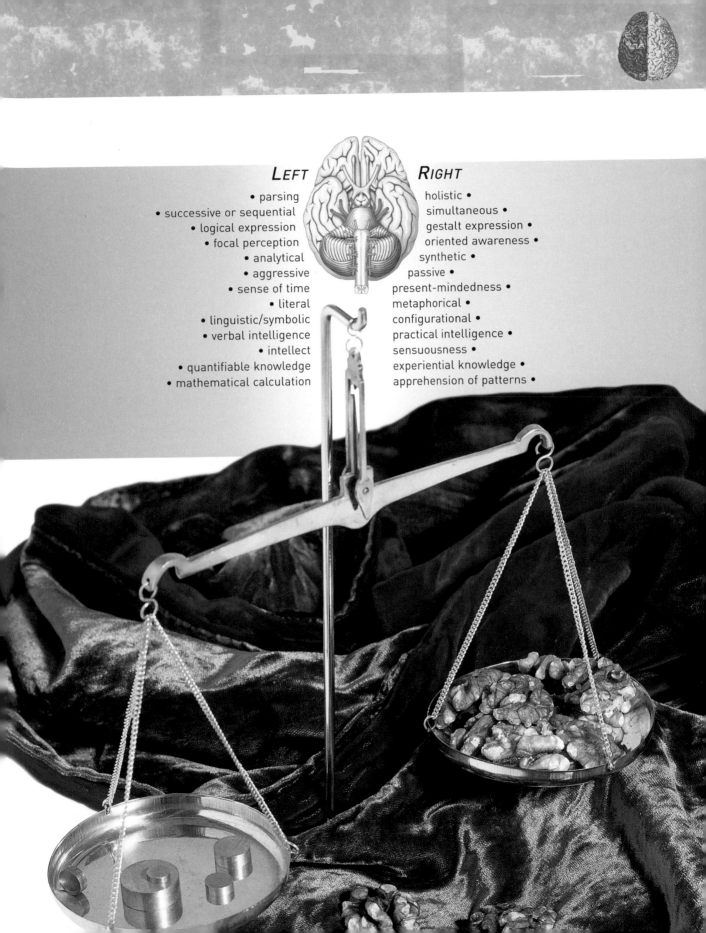

LEFT **RIGHT**

Left	Right
• parsing	holistic •
• successive or sequential	simultaneous •
• logical expression	gestalt expression •
• focal perception	oriented awareness •
• analytical	synthetic •
• aggressive	passive •
• sense of time	present-mindedness •
• literal	metaphorical •
• linguistic/symbolic	configurational •
• verbal intelligence	practical intelligence •
• intellect	sensuousness •
• quantifiable knowledge	experiential knowledge •
• mathematical calculation	apprehension of patterns •

Hemispheres of the Brain

THE ADVANTAGE OF DUAL PROCESSING

By specializing functions in different areas, the mental capacity of the brain is increased. Each hemisphere tends to analyze its own input first, only exchanging information with the other half when a considerable amount of processing has already taken place. With both hemispheres capable of working independently, human beings are able to process two streams of information at once. The brain then compares and integrates the information to obtain a broader and more in-depth understanding of a concept or object.

↑ **THE CORPUS CALLOSUM** is the bridge that enables the exchange of information between the two brain hemispheres.

→ **A PIANIST IS ABLE TO USE** both hands simultaneously to produce rhythm and melody.

Medical research, begun in the 1950s, discovered that the human brain has two hemispheres bridged by the corpus callosum. This bridge is the communication network between the two hemispheres. It was not until the 1960s that we began to understand the hemispheres' functions, when a team of psychologists and neurosurgeons led by Roger Sperry, Joseph Bogen, and Michael Gazzanniga began a series of pioneering experiments that seemed to indicate certain types of thinking were related to certain parts of the brain.

GROUNDBREAKING DISCOVERY

Sperry's team studied patients whose corpus callosum had been cut in an effort to reduce the severity of their epileptic attacks. While the epilepsy was largely cured, patients started to exhibit strange side effects—the two hemispheres seemed to behave differently. The left side seemed responsible for language and logical thought, while the right held sway over artistic abilities, intuition, and creativity.

In 1981 Dr. Sperry shared the Nobel Prize in medicine for his split-brain research, which serves as the basis for our current understanding of cerebral specialization in the human brain. His achievements are of major significance for developmental neurobiology and psychobiology.

Hemispheres of the Brain

THE KEY/CASE EXPERIMENT

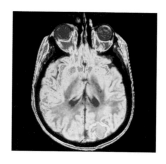

This experiment, carried out by Dr. Sperry, sets out to reveal that the two brain hemispheres have different ways of processing information.

A subject whose hemispheres have been surgically separated is seated in front of a screen that shows an image projected from the rear. The subject is told to focus his eyes on a dot painted in the center of the screen. On the left side of the screen, the letters K-E-Y are flashed for a split second. On the right side of the screen, the letters C-A-S-E are flashed at the same time. As the optic nerve connections from each eye cross to the opposite hemisphere of the brain, the messages shown on the screen are processed by the hemisphere of the brain opposite to the visual field on the screen.

When asked what had appeared on the screen, the subject replied that he saw the word CASE. This showed that the left hemisphere is specialized for processing verbal language in a sequential fashion. In other words, it recognized C-A-S-E, one letter after another, one word after another, and then one sentence after another. These letters had, to the left hemisphere, produced a sequential train of thought on a sequential line of reasoning.

However, when the same split-brain subject was asked to reach into a shelf under the screen with his left hand (the hand connected to the right side of the brain, and the side that saw the letters K-E-Y) and select one of a number of objects just by touch, the object he pulled out was a key.

This response revealed that the right side of the brain does not put incoming sensory information into sequential order and, therefore, does not articulate in speech what it knows it has seen. Instead, the right hemisphere processes information holistically (all at once), rather than parsing the information (breaking it up into small pieces).

→ **THE WORD CASE** is seen on the right and the word KEY is seen on the left.

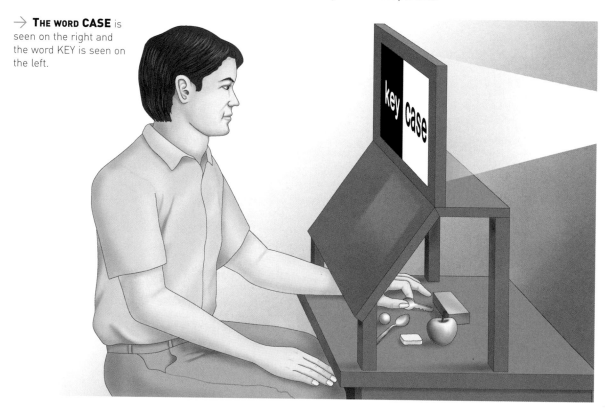

KEY

CASE

↑ **THE LEFT HEMISPHERE** sees the word CASE, the right hemisphere recognizes the object, the KEY.

Hemispheres of the Brain

THE SPOON EXPERIMENT

The second experiment with the same split-brain subject, carried out by Dr. Sperry, shows the capacity of the right hemisphere to think in visual-to-visual ways, in the same way that an artist will produce a finished product from initial sketches made in a notebook.

As in the KEY/CASE experiment, the subject is shown an image flashed onto a screen from the rear. This time, no letters or words are shown, instead the image of a spoon is flashed on the left side of the screen and, therefore, to the right side of the brain.

In this experiment the subject was unable to say linguistically what he had seen, although he clearly knew he had seen something. When, however, he reached into the same hidden shelf beneath the screen with his left hand, the subject selected the spoon, but this could not be named until it was withdrawn from the hidden shelf and seen by the

↓ **BELOW, THE SUBJECT REACHES UNDER THE SCREEN** to select the object he has seen but is unable to name.

left hemisphere. Therefore, it can be said that when the corpus callosum is severed, communication between the two hemispheres is stopped. It is then possible to determine the functions of the separate hemispheres by tests. The corpus callosum enables both hemispheres to work together for almost every activity, although one hemisphere or the other will predominate for a specific task.

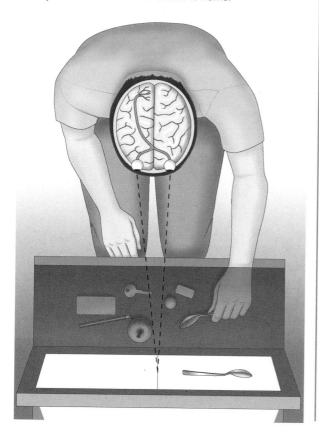

The left hemisphere of the brain is the rational, or logical side, and its main function is the ability to translate perceptions into logical and phonetic representations of reality, and to communicate with the outside world on the basis of this logical analysis. Concerned with reading, writing, counting, and computing, it processes information bit by bit and in a linear, logical fashion. This is called linear, vertical, or habitual thinking. The left hemisphere runs most of our systems and institutions, such as the military and government.

On the other hand, the right hemisphere is the intuitive hemisphere that imagines and perceives things holistically. It is appreciative of art and music, and is the part of our brain that explores worship, rituals, and mysticism, and our feelings. It is this side of the brain that reconstructs a whole pattern out of individual pieces, at the same time giving rise to new ideas and concepts. This is called holistic, lateral, intuitive, or metaphorical thinking, and it is highly

↑ **THE ANALYTICAL LEFT BRAIN** translates perceptions into logical and phonetic representations of reality.

specialized in the overall grasping of complex relationships, patterns, and structures.

The importance to each one of us of accessing both hemispheres of the brain is immense. The left brain learns in a conscious, methodical way. The right brain learns in a subconscious, creative, intuitive way. This means that to support the whole brain, function, logic, and intuition are equally important.

To SUMMARIZE:

Left brain	Right brain
• conscious thought	subconscious thought •
• logical analysis	emotional reaction •
• outer awareness	inner awareness •
• use of language	use of intuition •
• methods, rules	creativity •

← **THE RIGHT HEMISPHERE** controls subconscious thought and inner awareness, and is the part of the brain that explores worship, rituals, and mysticism.

The Enigma of the Brain

"In the matter of its own special activities, the brain is usually undisciplined and unreliable. We never know what it will do next."

Arnold Bennett, early twentieth-century novelist

In the last century, medical research has uncovered many mysteries of the "unreliable" human brain. To appreciate the concept of laterality, it is necessary to look at the physiology of the brain in more detail.

The brain is the part of the central nervous system that, in vertebrates, is contained within the skull. The human brain is a mass of pinkish gray tissue composed of about 10 billion nerve cells, each linked to another, and together responsible for the control of all mental functions.

The human brain is the supreme achievement of natural selection. Infinitely more complex than any computer, it is the natural product of hundreds of thousands of years of evolution. With an average weight of only 3 lb (1.36 kg), this web of nerves somehow manages to regulate all the body's systems, and at the same time absorbs and learns from a continual intake of thoughts, feelings, and memories.

↓ **WEIGHING 3 LB, THE BRAIN** is more complex than any computer, yet we utilize only 3 percent of its capacity.

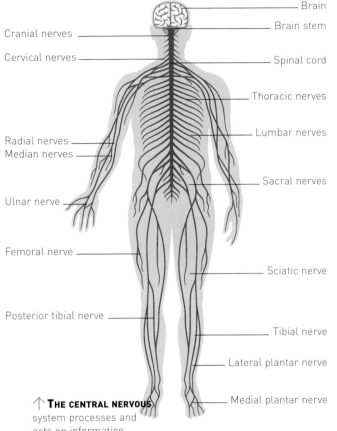

Brain

Brain stem

Cranial nerves

Cervical nerves

Spinal cord

Thoracic nerves

Lumbar nerves

Radial nerves

Median nerves

Sacral nerves

Ulnar nerve

Femoral nerve

Sciatic nerve

Posterior tibial nerve

Tibial nerve

Lateral plantar nerve

Medial plantar nerve

↑ **THE CENTRAL NERVOUS**
system processes and
acts on information
that is received from
the peripheral
nervous system.

→ **THE BRAIN IS PART**
of the central nervous
system that coordinates
and controls
bodily functions.

The brain is the control center for movement, sleep, hunger, and thirst—in fact, for virtually every activity necessary for survival. Additionally all the emotions—such as love, hate, elation, aggression, and fear—are controlled by the brain. It also receives and interprets countless signals sent to it from other parts of the body and from the external environment.

Despite centuries of thought, analysis, and research, many aspects of the workings of the brain remain an enigma. The more we learn about it, the more we realize how little we know.

LOOKING CLOSER AT THE BRAIN

The human brain consists of three major components. These are the large, dome-shaped cerebrum at the top; the smaller, and somewhat spherical, cerebellum at the lower right; and the central brain stem. Prominent in the brain stem are the medulla oblongata, shown as an egg-shaped enlargement in the center, and the thalamus, which lies between the medulla and the cerebrum. It is the cerebrum that is responsible for intelligence and reasoning, and helps to maintain balance and posture. The medulla is involved in involuntary functions such as respiration, while the thalamus is the relay center for electrical impulses traveling to and from the cerebral cortex.

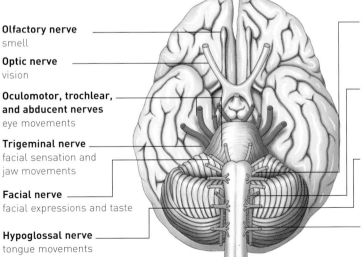

Olfactory nerve
smell

Optic nerve
vision

Oculomotor, trochlear, and abducent nerves
eye movements

Trigeminal nerve
facial sensation and
jaw movements

Facial nerve
facial expressions and taste

Hypoglossal nerve
tongue movements

**Acoustic
(vestibulocochlear)
nerve**
hearing and balance

**Glossopharyngeal
nerve**
taste and throat
sensations

Vagus nerve
breathing,
circulation, and
digestion

**Spinal accessory
nerve**
movement of neck
and back muscles

The Enigma of the Brain

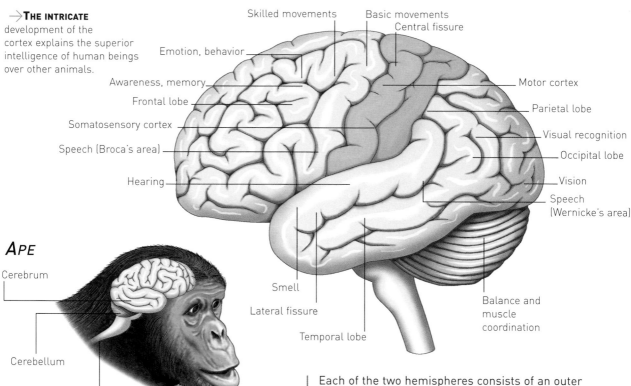

→ **THE INTRICATE** development of the cortex explains the superior intelligence of human beings over other animals.

Skilled movements

Basic movements

Central fissure

Emotion, behavior

Awareness, memory

Frontal lobe

Somatosensory cortex

Speech (Broca's area)

Hearing

Motor cortex

Parietal lobe

Visual recognition

Occipital lobe

Vision

Speech (Wernicke's area)

APE

Cerebrum

Cerebellum

Brain stem

Smell

Lateral fissure

Temporal lobe

Balance and muscle coordination

THE CEREBRUM

The largest part of the human brain is the cerebrum, which makes up approximately 85 percent of the brain's weight and its large surface area—the cortex—and intricate development, which accounts for the superior intelligence of human beings as opposed to other animals.

The cerebrum is divided by a central vertical fissure into right- and left-mirror image hemispheres. The corpus callosum is a mass of white nerve fibers that connects these two hemispheres and transfers information from one side to the other.

Within each hemisphere are fluid-filled spaces called ventricles that connect with a third ventricle, called the foramen of Monro. This is located between the hemispheres and, in turn, leads to a fourth ventricle, located in front of the medulla and the cerebellum, by means of a slender canal called the aqueduct of Sylvius.

Each of the two hemispheres consists of an outer layer of gray matter called the cerebral cortex, which is about ⅛ to ³⁄₁₆ inches (3 to 4 mm) thick. Fibers connect the cerebrum with other parts of the brain, the front of the brain to the back, different areas on the same side of the cerebrum, and one side of the brain to the other. Each hemisphere is divided into five lobes—the frontal, parietal, temporal, occipital, all located externally, and the insula, located internally.

THE CEREBELLUM

The cerebellum lies in the back part of the cranium beneath the cerebral hemispheres and is also composed of gray cells. It is made up of two hemispheres that are connected by white fibers called the vermis, which connect the cerebellum to other parts of the brain.

The cerebellum is essential to the control of movement and acts as a reflex center for the coordination and maintenance of equilibrium. This is the part of the brain that is responsible for all motor activity, from a tennis player smashing an ace into an opponent's court to a neurosurgeon performing the most delicate operation.

THE BRAIN STEM

The brain stem, shown here in cross-section, is divided into several parts, and is the lowest part of the brain. Its basic function is to serve as the path for messages traveling between the upper brain and the spinal cord, and it is also the basis of vital functions, such as breathing and heart rate, as well as reflex functions, such as eye movement.

← IN A DELICATE operation, the surgeon must utilize both hemispheres to coordinate knowledge and dexterity.

→ THE BRAIN STEM is the seat of basic and vital functions, such as breathing, blood pressure, and heart rate.

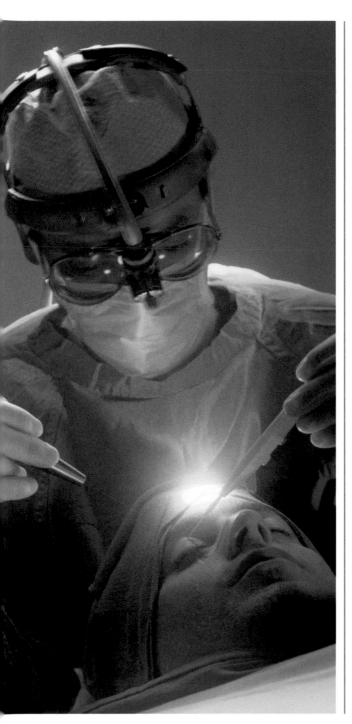

Thalamus
Hypothalamus
Midbrain
Pituitary
Pons
Medulla
Reticular formation

The brain stem consists of three main parts: the medulla, pons, and midbrain. Also distributed along the length of the brain stem is a network of cells, known as the reticular formation, that regulates the state of alertness. Components of the brain stem are necessary for performing most of the vital functions that are key to our survival.

Among these are the thalamus, the crucial relay station for incoming sensory signals and outgoing motor signals passing to and from the cerebral cortex; the hypothalamus, which regulates eating, drinking, sleep, body temperature regulation, and emotional behavior; and the medulla, which contains the vital control centers for cardiac and respiratory functions, as well as other reflex activities, such as vomiting.

→ IN ADDITION TO DEEP concentration, a top golfer needs both visual and muscular coordination.

The Enigma of the Brain

Many motor and sensory functions have been mapped to specific areas of the cerebral cortex. In general, these areas exist in both sides of the cerebrum, each serving the opposite side of the body.

The cerebral cortex is subdivided into several functional areas. The somomotor area, for example, is located just in front of the central fissure and is responsible for almost all body muscle movement.

Behind the central fissure is the somosensory area, which receives impulses from the skin's surface and below, and processes sensations, such as taste and touch. At the top of this area are the nerve cells for sensation from the toes, and at the base are those for the face. The area concerned with hearing, known as the auditory area, is in the upper temporal lobe; the area for seeing, the visual cortex, is in the back portion, or occipital lobe; and the area for smell, the olfactory area, is at the front internal portion.

Two areas that govern our language and speech have also been pinpointed to the cortex. These are Wernicke's area, concerned with the comprehension

→ **THE SINGLE-MINDED** determination of a top runner is a left-brain hemisphere function.

↓ **THE AUDITORY AREA,** which enables us to hear, is situated in the upper temporal lobe.

of spoken language, and Broca's area, responsible for the muscle movements of the throat and mouth used in speaking. A large area at the front of the cortex is used for awareness, intelligence, and memory.

↑ **OPERA SINGERS** require left-brain hemisphere attention to detail. Jazz singers, on the other hand, improvise with right-brain creativity.

The two hemispheres of the cortex are highly specialized and serve opposite sides of the body. For example, a touch on the right side of the body is perceived by the left somatosensory area, and to move the right arm, neurons in the left motor cortex have to be activated. In most individuals, the left hemisphere is dominant, which explains why a majority of people are right-handed.

How to Use This Book

SECTION ONE

There are a number of tests in section one, each designed to address a different facet of your brain. It is important that you attempt every one of these tests, otherwise you will not build a complete picture of your capacities. The answers to these tests are on pages 60 to 65, and scores will be allocated to your answers. The scores from each of these tests should be entered in the table provided on page 69, and the results plotted on the graph on the same page. This will show whether you have a left- or right-brain bias.

There are a number of psychometric or personality tests in this section. There is no right or wrong way to respond to these questions; simply answer intuitively and truthfully. Your only guidelines are the time limits that certain tests may impose.

The end of section one explains the tendencies your answers reveal, and outlines the areas of each side of the brain you need to work on. For example, you may find that you have a general left-brain bias, but that you still need to work on your verbal intelligence. Or that, despite having a good right-brain apprehension of patterns and spatial awareness, you could still improve your creative writing skills. Whatever level we attain at any skill or task, there is always room for improvement.

Example

Time limit

Test

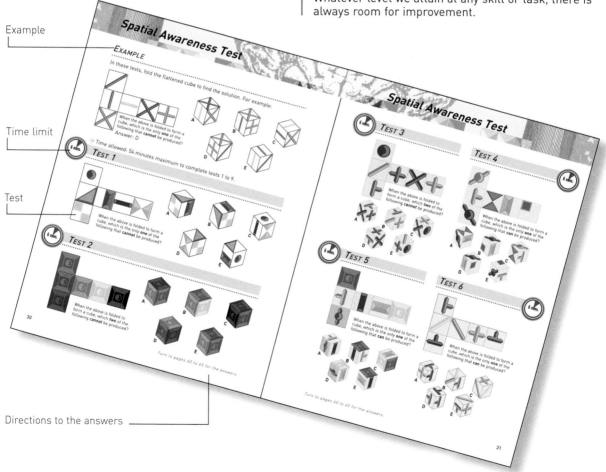

Directions to the answers

SECTION TWO

The workout in section two does exactly what it says —exercises your brain to get it into peak condition. This is achieved through a series of exercises and puzzles grouped into sections according to the function being trained. To be of greatest use to the reader, this book is designed so that all exercises to improve left-brain functions are on the left-hand page, and all exercises to improve the right-brain functions are on the right-hand page.

Using the knowledge of your own brain, gleaned from your personal assessment, hone those areas of the left or right side that need the most work. To attain peak condition and guarantee a truly balanced approach, it is recommended that you work through all the exercises, both left and right.

Chart to assess your brain bias

Exercises for those with a right-brain bias

Exercises for those with a left-brain bias

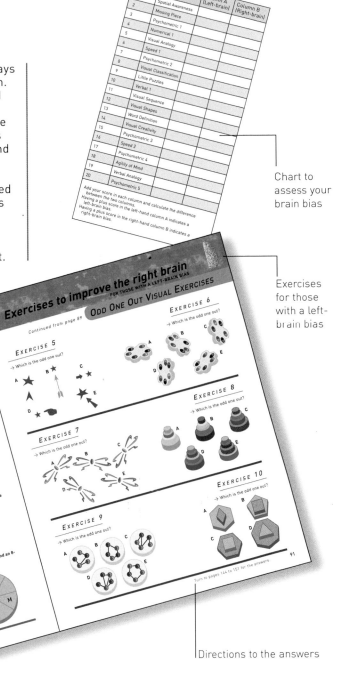

Directions to the answers

Number	Test description	Column A (Left-brain)	Column B (Right-brain)
1	Spatial Awareness		
2	Missing Piece		
3	Psychometric 1		
4	Numerical 1		
5	Visual Analogy		
6	Speed 1		
7	Psychometric 2		
8	Visual Classification		
9	Little Puzzles		
10	Verbal 1		
11	Visual Sequence		
12	Visual Shapes		
13	Word Definition		
14	Visual Creativity		
15	Psychometric 3		
16	Speed 2		
17	Psychometric 4		
18	Agility of Mind		
19	Verbal Analogy		
20	Psychometric 5		

Add your score in each column and calculate the difference between the two columns. Having a plus score in the left-hand column A indicates a left-brain bias. Having a plus score in the right-hand column B indicates a right-brain bias.

TESTS TO DETERMINE YOUR

BRAIN
BIAS

Complete the following tests
within the allocated time to determine
which side of your brain is dominant.

SECTION 1

Psychometric Testing

An individual's personality is the deeply ingrained, and usually enduring, patterns of their thoughts, feelings, and behaviors—and that combination is unique to every one of us. Personality implies predictability in how a person will act or react under different circumstances. Psychometric tests have existed since the turn of the nineteenth century, and in the past 25 to 30 years they have been brought into widespread use in industry and commerce. There are two main types of psychometric tests: personality questionnaires and aptitude tests.

The British Psychological Society's definition of a psychometric test is: "... any procedure on the basis of which inferences can be made concerning a person's capacity, propensity, or liability to act, react, experience, or to structure or order thought or behavior in particular ways."

While personality questionnaires are referred to as tests, this is quite misleading, because they do not have pass or fail scores, and are designed to measure attitudes, habits, and values. Usually, these tests are not timed. The questionnaires can be incorporated into a company's application form, or used at the second stage of a selection procedure.

BE TRUE TO YOURSELF

When faced with a personality questionnaire, simply follow the instructions and answer each question honestly. When in an interview situation, avoid answering what you think the prospective employer wants to hear, since this is likely to be spotted when the results are being analyzed, or could result in you being offered a job that does not suit you very well.

Another form of psychometric testing is aptitude tests. These are designed to give an objective assessment of the candidate's ability in verbal understanding, numeracy, and diagrammatic, or spatial, reasoning. These particular tests are often timed and marked, and usually have a cutoff point below which you either have failed or need to be reassessed. However, this is not the specific purpose of any of the tests in section one. Their purpose is simply to determine laterality—to show whether you are dominated by the right or left side of your brain.

Many factors determine which side is dominant in your brain, and your personality is often a good guide. The five psychometric tests in section one explore a range of aspects of your character and make-up. There is no need to read them through first—just answer them intuitively, and without too much imagination or forethought. There is no right or wrong response.

↑→ PSYCHOMETRIC TESTING is being used more often by prospective employers alongside traditional interviewing techniques. It is vital to answer each question with complete honesty when completing any personality questionnaire.

← A RIGHT-HEMISPHERE person will find it difficult to concentrate fully on any one subject.

Spatial Awareness Test

EXAMPLE

In these tests, fold the flattened cube to find the solution. For example:

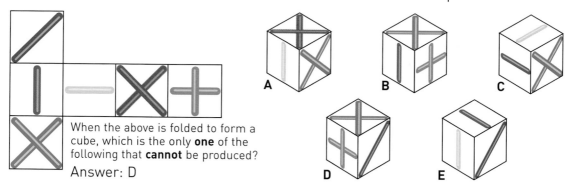

When the above is folded to form a cube, which is the only **one** of the following that **cannot** be produced?

Answer: D

→ Time allowed: 54 minutes maximum to complete tests 1 to 9.

 6 mins.

TEST 1

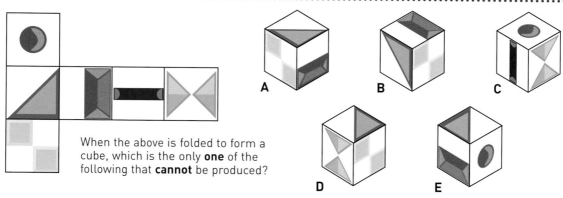

When the above is folded to form a cube, which is the only **one** of the following that **cannot** be produced?

 6 mins.

TEST 2

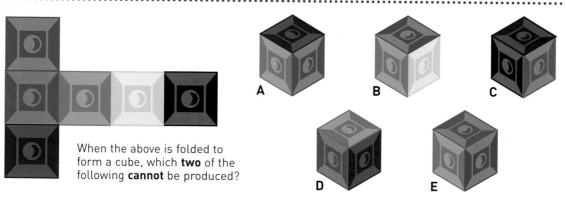

When the above is folded to form a cube, which **two** of the following **cannot** be produced?

Spatial Awareness Test

TEST 3

When the above is folded to form a cube, which **two** of the following **cannot** be produced?

A   B   C   D   E

TEST 4

When the above is folded to form a cube, which is the only **one** of the following that **can** be produced?

A   B   C   D   E

TEST 5

When the above is folded to form a cube, which is the only **one** of the following that **can** be produced?

A   B   C   D   E

TEST 6

When the above is folded to form a cube, which is the only **one** of the following that **can** be produced?

A   B   C   D   E

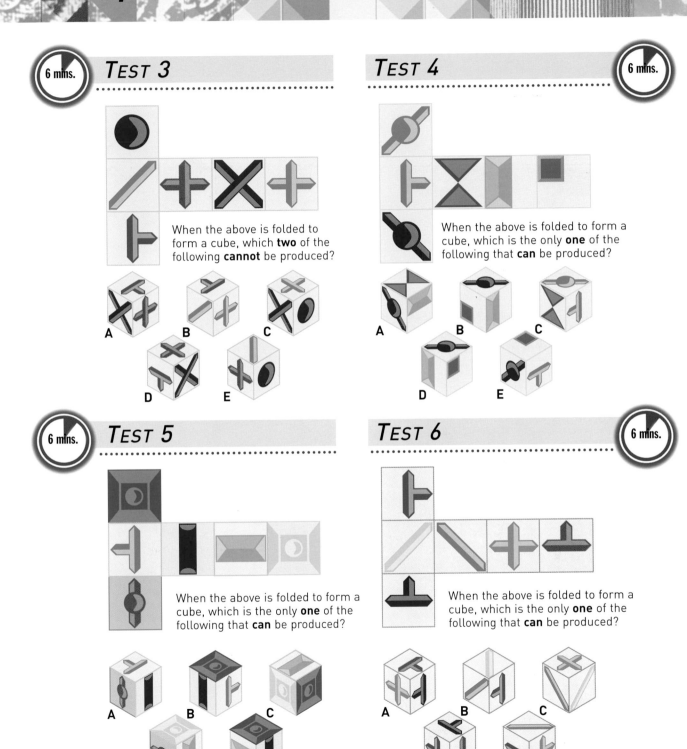

Turn to pages 60 to 65 for the answers.

Spatial Awareness Test

6 mins. ## TEST 7

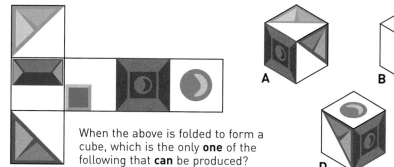

When the above is folded to form a cube, which is the only **one** of the following that **can** be produced?

6 mins. ## TEST 8

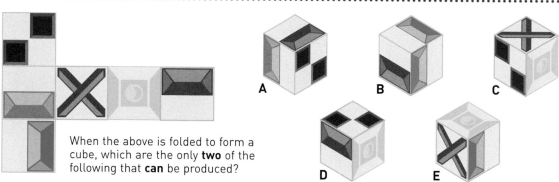

When the above is folded to form a cube, which are the only **two** of the following that **can** be produced?

6 mins. ## TEST 9

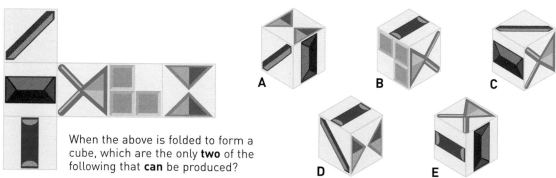

When the above is folded to form a cube, which are the only **two** of the following that **can** be produced?

Turn to pages 60 to 65 for the answers.

Missing Piece Test

EXAMPLE

In each of the following tests, find the missing piece that should fit into the grid to complete the array correctly. For example:

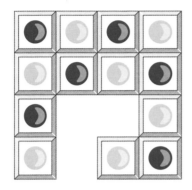

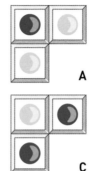

A

B

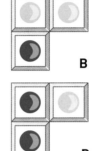

C

D

Answer:
C. It completes the pattern of alternate red and yellow circles.

→ Time allowed: 60 minutes maximum to complete tests 1 to 10.

 6 mins.

TEST 1

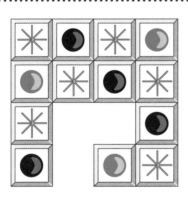

A   B

C   D

TEST 2

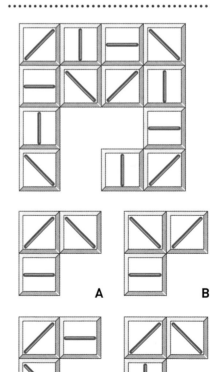

6 mins.

A   B

C   D

TEST 3

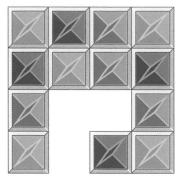

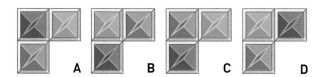

TEST 4

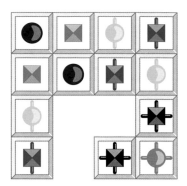

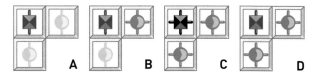

TEST 5

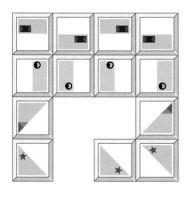

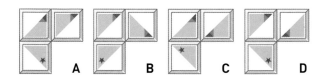

TEST 6

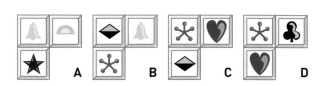

Turn to pages 60 to 65 for the answers.

Missing Piece Test

Turn to pages 60 to 65 for the answers.

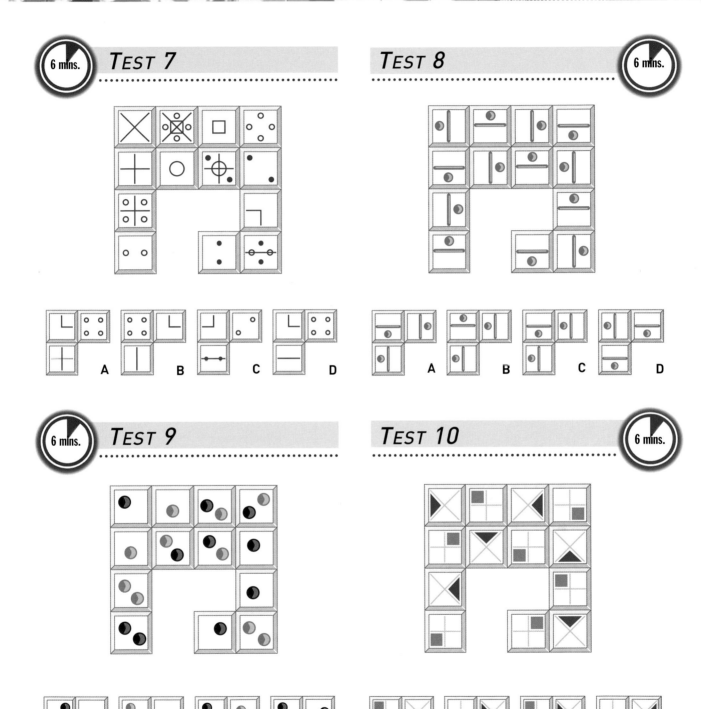

TEST 7   6 mins.

TEST 8   6 mins.

TEST 9   6 mins.

TEST 10   6 mins.

Psychometric Test 1

Choose the one option out of the three provided in each question that you think is most applicable to yourself.

→ Time allowed: 15 minutes maximum to complete questions 1 to 15.

1. You feel most confident when:
a. making decisions with a partner
b. taking instructions from an expert
c. making your own decisions

2. Which of the following words most accurately describes you?
a. flexible
b. forgiving
c. forceful

3. Which of the following words appeals to you the most?
a. free
b. fair
c. famous

4. Which of the following do you admire the most?
a. a great construction, such as the Taj Mahal
b. a thing of natural beauty, such as the Great Barrier Reef
c. a great operatic voice, such as Pavarotti

5. Which of these amusement park attractions do you prefer?
a. bumper cars
b. pinball machines
c. the haunted house

6. You are asked to produce your birth certificate. How long would it take you to find it?
a. less than five minutes
b. more than one hour
c. less than one hour

7. Which of the following words most accurately describes you?
a. intellectual
b. impulsive
c. intuitive

8. Which of these annoys you the most?
a. untruthfulness
b. intolerance
c. insolence

9. How would you best like to be described?
a. an upright, law-abiding citizen
b. one of life's great characters
c. a great innovator

10. Which aspect of gardening interests you the most?
a. garden design
b. the opportunity to work with the land
c. the cultivation of plants and flowers

11. Which of the following most accurately represents your views on global warming?
a. nature will take its course whatever
b. I do not believe global warming is the problem it is made out to be
c. we must do something about it before it is too late

12. You go to a new restaurant for a three-course celebration meal. What would choose?
a. something you have never tried before for the first course
b. something new, but highly recommended, for the main course
c. something you know you like to eat

13. Which of the following best describes you?
a. living by the rules
b. something of a rebel
c. live and let live

14. Which one of the following do you most admire?
a. Stephen Hawking
b. Margaret Thatcher
c. Bill Gates

15. Which of the following most fascinated you as a child?
a. jigsaw puzzles
b. comic books
c. kaleidoscopes

Turn to pages 60 to 65 for the answers.

Numerical Test

EXAMPLE

In each of the following, work out the sequence and insert the correct number to replace the question mark. For example:

0, 1, 3, 6, 10, ?   Answer: 15

→ Time allowed: 40 minutes maximum to complete questions A to S.

a 100, 101, 103, 107, 115, 122, ?

b 1, 2, 4, 3, 7, 4, ?

c 135, 791, 113, 151, 719, ?

d 1, 2, 4, 7, 8, 10, ?

e 100, 99, 96.5, 92.5, ?

f 10, 10, 11.5, 15, 13, 22.5, 14.5, ?

g 1, 3, 7, 15, 31, ?

h 0, 4, 4, 8, 12, 20, ?

i 0, 1, -1, 0, -2, ?

j 85, 95, 91, 89, 97, 83, 103, ?

k 5, 4.5, 13.5, 10.5, 3.5, ?

l 10, 11, 12.1, 13.31, ?

m 18, 9, 10, 2, 4, -3, ?

n 25, 5, 10, 2, 7, ?

o 1, 3, 5, 15, 17, ?

p 100, 80, 70, 65, ?

q 16, 17, 21, 30, 46, ?

r 79, 88, 96, 102, 104, ?

s 12, 30, 75, ?

Turn to pages 60 to 65 for the answers.

Visual Analogy Test

EXAMPLE

Choose the image that mirrors the example set in each test. For example:

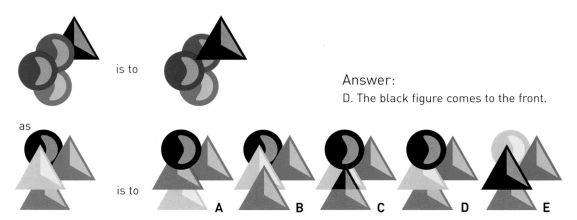

Answer:
D. The black figure comes to the front.

→ Time allowed: 27 minutes maximum to complete tests 1 to 9.

 TEST 1

 TEST 2

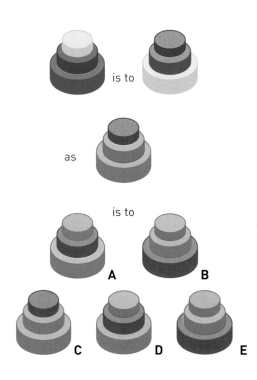

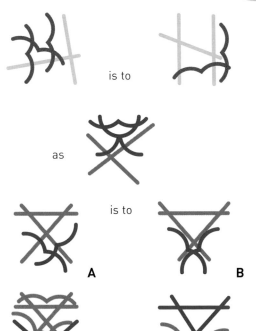

Turn to pages 60 to 65 for the answers.

Visual Analogy Test

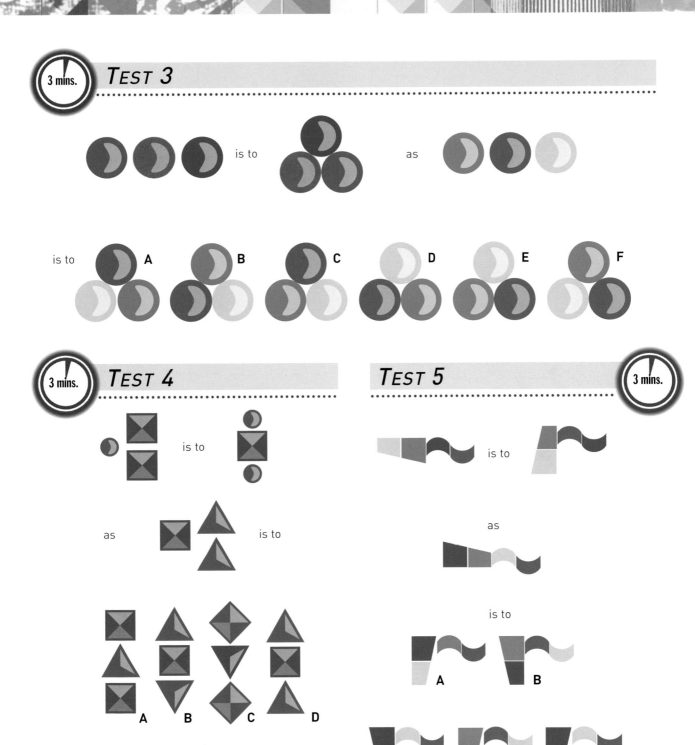

Turn to pages 60 to 65 for the answers.

Visual Analogy Test

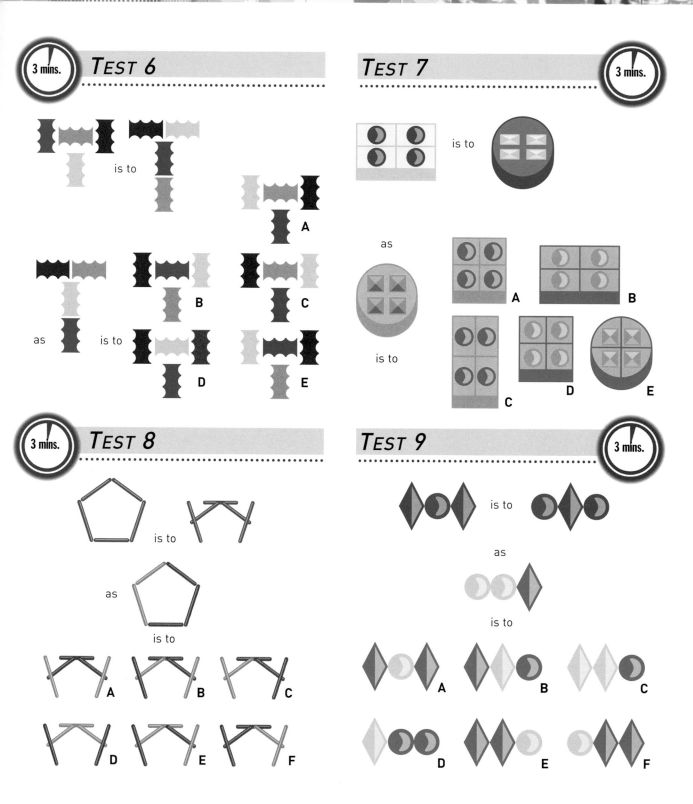

Turn to pages 60 to 65 for the answers.

Speed Test 1

EXAMPLE

....

Each statement is either TRUE or FALSE. For example:

49 minutes after 9:00 A.M. is the same as 11 minutes before 10:00 A.M.   Answer: TRUE

$\rightarrow$ Time allowed: 15 minutes maximum to complete questions 1 to 29.

....

1. The word MAGNETIC can be spelled using all but one of the letters from the word ENIGMATIC.

   TRUE   FALSE

2. If the word TOP is written under the word SAD and the word DIN is written below the word TOP, the word SIN is formed diagonally.

   TRUE   FALSE

3. If Friday is the third day of the month, then the following Tuesday is the seventh day of the month.

   TRUE   FALSE

4. Two of the following numbers add up to 19:

   3, 4, 8, 10, 13, 17

   TRUE   FALSE

5. This sentence contains the letter E eight times.

   TRUE   FALSE

6. The word GEOGRAPHY is spelled out by the first letters of the sentence: *George Eliot's old grand-mother rode a pig yesterday.*

   TRUE   FALSE

7. *Pull a bat, I hit a ball up* is a sentence that reads the same backward and forward.

   TRUE   FALSE

8. The odd numbers in the sequence 78432169 add up to 21.

   TRUE   FALSE

9. An upside-down clock's minute hand points to the left when it is quarter past one.

   TRUE   FALSE

10. In the alphabet, the letter one before the letter two after the letter F is the letter G.

    TRUE   FALSE

11. The number 717328162 written in reverse is 261823717.

    TRUE   FALSE

12. If my house is tenth from the end of the row and sixth from the other end, there are 17 houses in the row.

    TRUE   FALSE

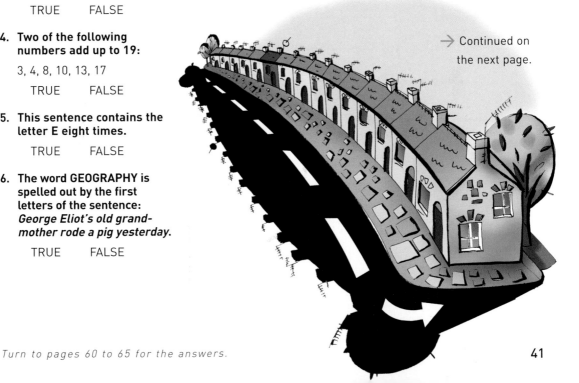

$\rightarrow$ Continued on the next page.

Turn to pages 60 to 65 for the answers.

13. Eight ostriches and six tigers have 40 legs.

    TRUE   FALSE

14. The number 4 is 20 less than when multiplied by 6 times itself.

    TRUE   FALSE

15. The middle two letters of the words CRAYON, AFFRAY, and SPIDER can be arranged to spell FRIDAY.

    TRUE   FALSE

16. If I walk three miles north, then two miles to my right, then two miles south, I will then have to walk two miles east to return to my starting point.

    TRUE   FALSE

17. If I wrote down all the numbers from 1 to 20, I would write down the number one 12 times.

    TRUE   FALSE

18. I have $100 and give away 40 percent, then another $25, after which I am left with just $45.

    TRUE   FALSE

19. A square measuring two inches on each side will fit into a circle having a radius of two inches.

    TRUE   FALSE

20. Tony is shorter than Alice and Ian is taller than Tony, therefore Tony is the shortest of the three.

    TRUE   FALSE

21. 1, 3, 7, 15, ?

    The number 31 is the next logical number in the sequence above.

    TRUE   FALSE

22. If three people all said hello to each other once, the word *hello* would be spoken six times.

    TRUE   FALSE

23. The sentence **He raced around** contains the name of a tree.

    TRUE   FALSE

24. I know three different routes from A to B and two different routes from B to C, therefore I know five different routes from A to C.

    TRUE   FALSE

25. If I looked in a mirror and saw a reflection of my digital clock showing 11 minutes past 12, the correct time would be 12 minutes past 11.

    TRUE   FALSE

26. Two of the numbers 18, 56, 32, 54, 24, 68 add up to 100.

    TRUE   FALSE

27. By drawing two lines, it is possible to divide a triangle up into three small triangles.

    TRUE   FALSE

28. If you have four-fifths of $100 and spend $22, you are left with $58.

    TRUE   FALSE

29. The fifteenth of December comes 35 days after the tenth of November.

    TRUE   FALSE

Turn to pages 60 to 65 for the answers.

Psychometric Test 2

Choose the one option out of the three provided in each question that you think is most applicable to yourself.

→ Time allowed: 15 minutes maximum to complete questions 1 to 15.

1. **Do you consider yourself ruthless?**
   a. no, I don't like people who are ruthless
   b. maybe a little
   c. I am when it comes to getting what I want

2. **Which of these historical characters would you most like to shake the hand of and congratulate?**
   a. Mother Teresa
   b. Robert the Bruce
   c. Jesse James

3. **If a rather loud verbal argument broke out between two colleagues at work, would you:**
   a. keep your head down and leave well enough alone
   b. perhaps try to calm things down if you thought it wise to do so in the circumstances
   c. almost always get involved with relish either by joining in or trying to calm things down

4. **What are your feelings on the old adage, *attack is the best form of defense*?**
   a. strongly disagree
   b. agree sometimes
   c. agree it is usually the best policy

5. **How often do you lose your temper?**
   a. never
   b. up to four times a year
   c. more than four times a year

6. **You are driving your car, and another driver almost causes you to be involved in an accident. Do you:**
   a. shrug your shoulders and consider yourself thankful that there wasn't an accident
   b. mutter about the other driver to yourself or a companion in the car
   c. gesticulate or shout at the other driver to let him know what you think of him

7. **Would you in any circumstances resort to violence?**
   a. never
   b. perhaps under certain circumstances
   c. yes

8. **How often do you use mild expletives?**
   a. never or occasionally
   b. quite often
   c. very often

9. **Have you ever walked along the street with your fists clenched?**
   a. I am not aware that I have
   b. occasionally, if I am feeling tense or upset
   c. yes

10. **What are your feelings on the following statement, *aggressive behavior is necessary sometimes as a means to an end*?**
    a. strongly disagree
    b. maybe agree in certain circumstances
    c. agree

11. **How often do you use strong expletives?**
    a. never
    b. occasionally
    c. more than occasionally

12. **You are in a cinema complex faced with a choice of three films. Which would you choose?**
    a. *The Sound of Music*
    b. *Patton*
    c. *Psycho*

13. **If someone did you a particularly bad deed, would you:**
    a. wonder why they had done it, but shrug it off
    b. have a word with them and ask them why
    c. try to get back at them

14. **How often do you question or analyze your own conduct?**
    a. frequently
    b. sometimes
    c. rarely

15. **How often have you vented your frustration on a store assistant who is not being very helpful?**
    a. very rarely or never
    b. occasionally
    c. more than occasionally

Turn to pages 60 to 65 for the answers.

Visual Classification Test

EXAMPLE

Look very carefully at the boxes in each test below, and find the common bond.
For example:

To which of the five boxes below can a dot be added so that it meets the same conditions as the box on the left?

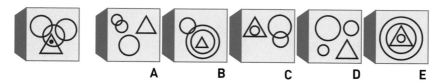

A   B   C   D   E

Answer: C. The dot appears in the triangle and one circle.

→ Time allowed: 27 minutes maximum to complete tests 1 to 9.

 3 mins.

TEST 1

 To which of the five boxes below can a dot be added so that it meets the same conditions as the box to the left?

A   B   C   D   E

TEST 2

 3 mins.

 Which box below has the most in common with the box to the left?

A   B   C   D   E

 3 mins.

TEST 3

 To which of the five boxes below can a dot be added so that both dots meet the same conditions as the box to the left?

A   B   C   D   E

TEST 4

3 mins.

 Which box below has the most in common with the box to the left?

A   B   C   D   E

Turn to pages 60 to 65 for the answers.

Visual Classification Test

 TEST 5

 To which of the five boxes below can a dot be added so that both dots then meet the same conditions as the box to the left?

A   **B**   **C**   **D**   **E**

 TEST 6

  To which of the five boxes below can a dot be added so that both dots then meet the same conditions as the box to the left?

A   **B**   **C**   **D**   **E**

 TEST 7

 To which of the five boxes below can a dot be added so that both dots then meet the same conditions as the box to the left?

A   **B**   **C**   **D**   **E**

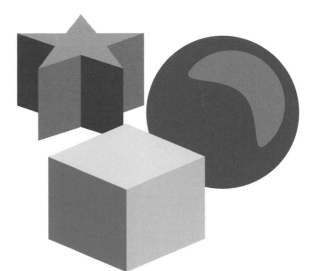

  TEST 8

Which box below has the most in common with the box to the left?

A   **B**   **C**   **D**   **E**

TEST 9

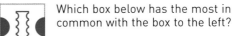 Which box below has the most in common with the box to the left?

A   **B**   **C**   **D**   **E**

Turn to pages 60 to 65 for the answers.

Little Puzzles Test

EXAMPLE

Which four numbers should replace the question marks?

Answer:
9	3
8	7

9	8	7	3	9
3	7	8	?	?
7	3	9	?	?
8	9	3	7	8
9	8	7	3	9

→ Time allowed: 56 minutes maximum to complete tests 1 to 14.

TEST 1

Which number should replace the question mark?

		13		
				25
	?			
			44	
51				

TEST 2

3829 is to 2398   as 5672 is to 7526,

so 1534 is to ?

TEST 3

What do the following names have in common?

Milford Haven   Switzerland
Ludwigshafen   Persian Gulf
Westmorland   Glastonbury
   Charleston

TEST 4

A man is alone on an island with no food and water and no form of transportation, yet he does not fear for his safety. Why?

TEST 5

What number should replace the question mark?

```
        5
      7   9
   10   4   8
  4   7   10   3
7   10   4   7   ?
```

TEST 6

Three years ago I visited Pasadena, two years ago my travels took me to Adelaide, and last year I spent some time, and all my money, in Las Vegas. Next year I plan to visit either Anglesey in Wales, Veracruz in Mexico, or the country of Thailand. Which one of these do you think I will visit?

TEST 7

What number should replace the question mark?

5 / 7 \ 3 / 6   9 / 19 \ ? / 10   4 / 12 \ 5 / 4

Turn to pages 60 to 65 for the answers.

Little Puzzles Test

TEST 8

4 mins.

What number should replace the question mark?

TEST 9

4 mins.

What symbols should replace the question marks?

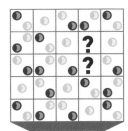

TEST 10

4 mins.

What number should replace the question mark?

7564928

4928756

8756492

?

TEST 11

4 mins.

What number should replace the question mark?

21.3, 21.9, 23.1, 23.7, 24.9, 26.4, **?**

TEST 12

4 mins.

I L A

What letter completes the above sequence?

TEST 13

4 mins.

What letter should replace the question mark?

TEST 14

4 mins.

  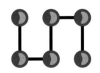 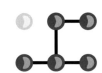

What is the next figure in the above sequence?

Turn to pages 60 to 65 for the answers.

Verbal Test

EXAMPLE

Use your vocabulary knowledge to complete this test. For example:

Which word in brackets is closest in meaning to the word in capitals?

CONTENTION (cognition, atrophy, trepidation, discord)   Answer: discord

→ Time allowed: 20 minutes maximum to complete questions 1 to 19.

20 mins.

1. **Which two words are most opposite in meaning?**

   irrational, pleasing, privy, wise, perceptible

2. **Which two words are closest in meaning?**

   change, stereotype, delete, pigeonhole, identify

3. **Which word in brackets is opposite in meaning to the word in capitals?**

   PLAUSIBLE (glib, incredible, banal, hostile)

4. **Which two words are closest in meaning?**

   incision, nostrum, vignette, panacea, progeny

5. **Which word in brackets is opposite in meaning to the word in capitals?**

   INVOLVED (planned, simple, volitional, secure)

6. **Which word in brackets is closest in meaning to the word in capitals?**

   HOST (individual, myriad, friend, specter)

7. **Which two words are most opposite in meaning?**

   erroneous, rambling, rousing, faultless, cultured

8. **Which word in brackets is most opposite to the word in capitals?**

   INVIOLATE (simple, sacred, distinct, polluted, sacrosanct)

9. **Which two words are closest in meaning?**

   gloss, verve, flaunt, zeal, languor, faith

10. **Which word in brackets is closest in meaning to the word in capitals?**

    EMPIRICAL (pragmatic, powerful, unequivocal, ardent, esteemed)

11. **Which word in brackets is most opposite to the word in capitals?**

    NEGATE (respect, confirm, construct, transcend)

12. **Which two words are closest in meaning?**

    acute, agog, involved, impatient, akin

13. **Which two words are most opposite in meaning?**

    submit, condone, encapsulate, plot, protract, censure

14. **Which word in brackets is closest in meaning to the word in capitals?**

    GLUTINOUS (abundant, saturnine, viscid, voracious, pallid)

15. **Which two words are most opposite in meaning?**

    autocratic, piercing, mellifluous, painless, hostile, ephemeral

16. **Which word in brackets is most opposite to the word in capitals?**

    SLY (unbigoted, priggish, fair, canny, ingenuous)

17. **Which word in brackets is closest in meaning to the word in capitals?**

    FRACTIOUS (resilient, petulant, wholesome, frantic, persistent)

18. **Which word in brackets is most opposite to the word in capitals?**

    VITAL (destructive, redundant, weak, malignant)

19. **Which two words are most opposite in meaning?**

    pure, regal, rosy, curt, wan, fine

Turn to pages 60 to 65 for the answers.

Visual Sequence Test

EXAMPLE

In the following tests, choose the image that is next in the sequence.
For example:

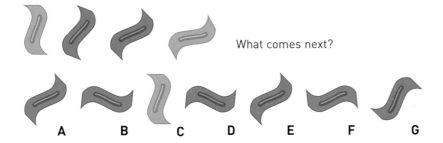

What comes next?

Answer:
B. The figure is tumbling over and the color sequence is pink, blue, brown, repeated.

→ Time allowed: 27 minutes maximum to complete tests 1 to 9.

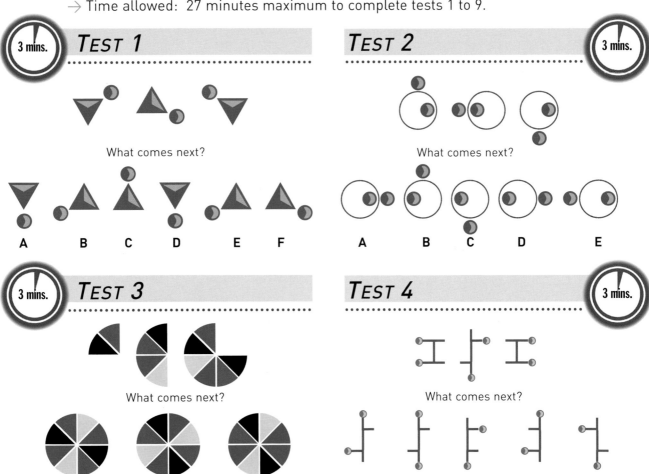

3 mins.

TEST 1

What comes next?

A B C D E F

3 mins.

TEST 2

What comes next?

A B C D E

3 mins.

TEST 3

What comes next?

A B C

TEST 4

3 mins.

What comes next?

A B C D E

Turn to pages 60 to 65 for the answers.

Visual Sequence Test

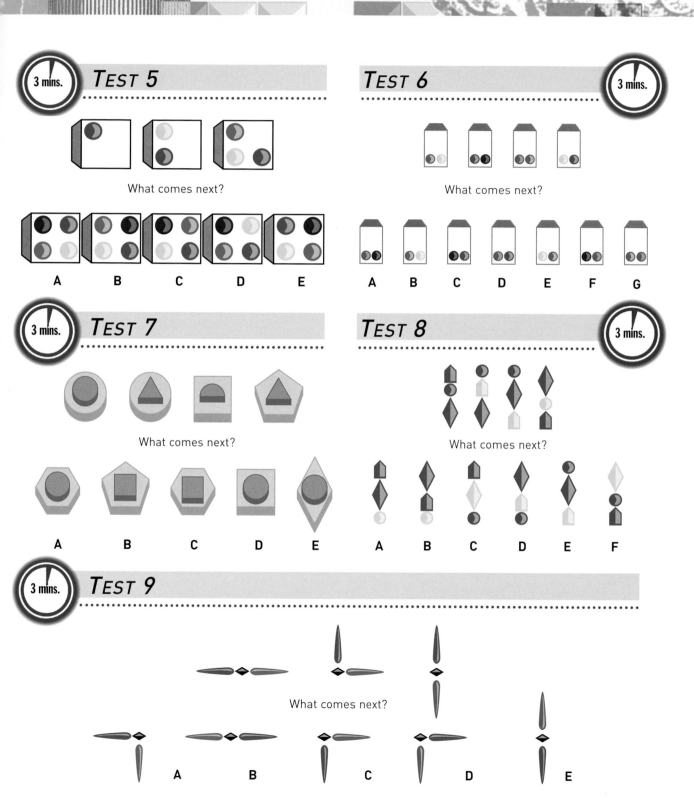

TEST 5

What comes next?

A   B   C   D   E

TEST 6

3 mins.

What comes next?

A   B   C   D   E   F   G

TEST 7

3 mins.

What comes next?

A   B   C   D   E

TEST 8

3 mins.

What comes next?

A   B   C   D   E   F

TEST 9

3 mins.

What comes next?

A   B   C   D   E

50

Turn to pages 60 to 65 for the answers.

Visual Shapes Test

EXAMPLE

In the following tests, fit the required number of pieces together to form the suggested shape. For example:

Which three of the four pieces to the right can be fitted together to form a perfect square?

Answer:

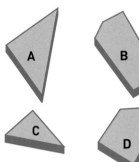

→ Time allowed: 54 minutes maximum to complete tests 1 to 9.

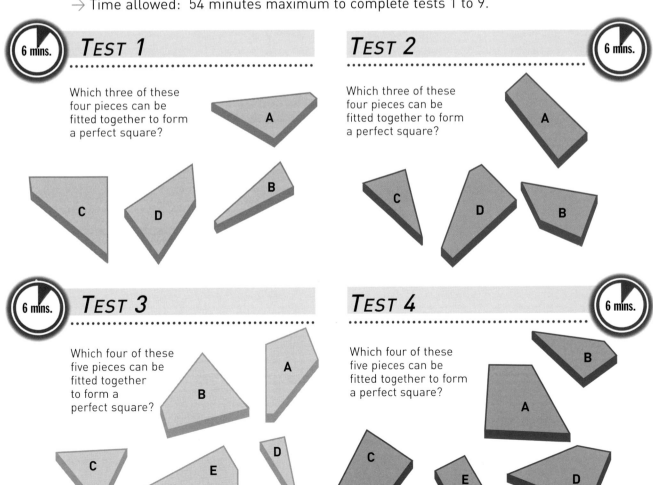

6 mins.

TEST 1

Which three of these four pieces can be fitted together to form a perfect square?

6 mins.

TEST 2

Which three of these four pieces can be fitted together to form a perfect square?

6 mins.

TEST 3

Which four of these five pieces can be fitted together to form a perfect square?

6 mins.

TEST 4

Which four of these five pieces can be fitted together to form a perfect square?

Turn to pages 60 to 65 for the answers.

Visual Shapes Test

6 mins. **TEST 5**

Which three of the four pieces below can be fitted together to form a circle?

6 mins. **TEST 6**

Which four of the five pieces below can be fitted together to form a circle?

6 mins. **TEST 7**

Which piece below, when fitted to the piece above, will form a perfect square?

6 mins. **TEST 8**

Which piece below, when fitted to the piece above, will form a perfect square?

6 mins. **TEST 9**

Which four of the five pieces below can be fitted together to form a cube?

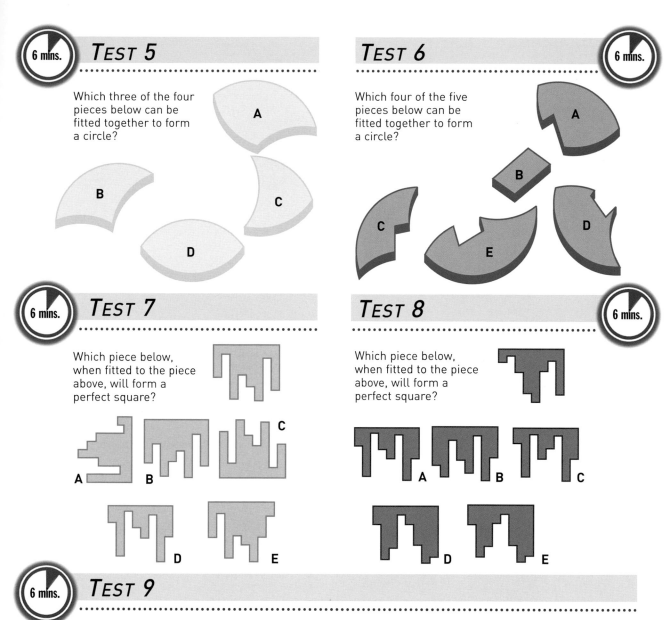

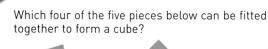

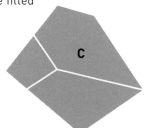

52   *Turn to pages 60 to 65 for the answers.*

Word Definition Test

EXAMPLE

In each of the following exercises, definitions of the key word are given. Choose the one correct definition.

For example:

saturnalia
a. gloom
b. passivism
c. revelry
d. strength

Answer:
    c. revelry

→ Time allowed: 15 minutes maximum to complete questions 1 to 19.

1. **saw**
   a. knowledge of appropriate behavior
   b. distinctive quality
   c. a hinged doorway
   d. a familiar old proverb

2. **ephemera**
   a. printed pamphlets
   b. outer layer of skin
   c. a band of followers
   d. descriptive language

3. **qualm**
   a. slight tremor
   b. musical note
   c. doubt
   d. inflammation

4. **rakish**
   a. charming
   b. crafty
   c. jaunty
   d. disreputable

5. **immiscible**
   a. confused
   b. unable to blend
   c. relating to fire
   d. inexcusable

6. **hypothetical**
   a. conjectural
   b. false
   c. unlikely
   d. symbolic

7. **facile**
   a. inappropriate
   b. artificial
   c. lively
   d. superficial

8. **truculent**
   a. loud
   b. bulging out
   c. aggressive
   d. restless

9. **wraith**
   a. young person
   b. ghost
   c. whimsy
   d. traveler

10. **grommet**
    a. thin mortar
    b. underground passage
    c. low wall or house
    d. ring or eyelet

11. **hallux**
    a. gaunt
    b. the big toe
    c. sixteenth-century axe
    d. small crest

12. **advocate**
    a. to recommend
    b. to admonish
    c. to praise
    d. to appeal

13. **veracity**
    a. excess of words
    b. expediency
    c. greed
    d. truthfulness

14. **serried**
    a. sequential
    b. tightly packed
    c. tangled
    d. having teeth, like a saw

15. **expatiate**
    a. elaborate
    b. speed up
    c. emigrate
    d. consume

16. **luminary**
    a. emission of light
    b. relating to the moon
    c. crescent-shaped space or opening
    d. person of brilliant achievement

17. **gloaming**
    a. mist
    b. twilight
    c. riverbank
    d. heath land

18. **glutinous**
    a. offensive
    b. excessive
    c. sticky
    d. deep

19. **foment**
    a. to encourage or provoke
    b. to rummage
    c. to cook slowly
    d. to visit frequently

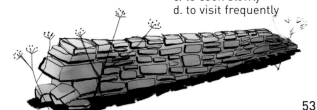

Turn to pages 60 to 65 for the answers.

Visual Creativity Test

EXAMPLE

Make one cut to divide the figure into two identical pieces.

→ Time allowed: 36 minutes maximum to complete tests 1 to 9.

TEST 1 *4 mins.*

Divide the square into four pieces of equal size and shape so that each of the four pieces contains one each of the four different colors.

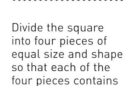

TEST 2 *4 mins.*

Move just three matchsticks to create four triangles.

TEST 3 *4 mins.*

Complete the route from A to B without taking your pencil from the paper and by traveling along every line. Lines may cross, but must not be retraced.

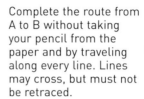

TEST 4 *4 mins.*

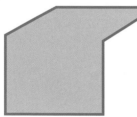

Make one cut of two straight lines that divides the figure into two halves of equal size and shape.

TEST 5 *4 mins.*

Which circle is directly opposite the circle that is three places counterclockwise away from the circle directly opposite the black circle?

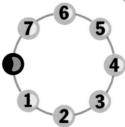

TEST 6 *4 mins.*

Four cubic blocks have been glued together and their outside faces have been painted so no two sections of the same color share adjoining edges. How many sections of each of the three colors appear in all on the entire cube?

TEST 7 *4 mins.*

Three views of the same color cube are shown. What color is opposite the red face?

54

Turn to pages 60 to 65 for the answers.

Psychometric Test 3

TEST 8

4 mins.

Divide this figure into two equal segments.

TEST 9

4 mins.

Which is the odd one out?

A

B

C

D

E

In each of the following, mark the one word that you believe is most applicable to yourself.

→ Time allowed: 20 minutes maximum to complete questions 1 to 20.

20 mins.

1. a. thoughtful
   b. stressed
   c. busy

2. a. practical
   b. inventive
   c. studious

3. a. bold
   b. accommodating
   c. submissive

4. a. discontented
   b. anxious
   c. relaxed

5. a. down-to-earth
   b. typical
   c. complex

6. a. visionary
   b. compliant
   c. businesslike

7. a. hedonistic
   b. thorough
   c. workaholic

8. a. undemanding
   b. content
   c. complex

9. a. concerned
   b. patient
   c. persistent

10. a. disordered
    b. careless
    c. methodical

11. a. calculating
    b. serene
    c. emotional

12. a. habitual
    b. unconventional
    c. orthodox

13. a. imaginative
    b. impatient
    c. careful

14. a. philosophical
    b. rational
    c. environmental

15. a. restrained
    b. volatile
    c. steady

16. a. accepting
    b. naive
    c. demanding

17. a. tidy
    b. chaotic
    c. analytical

18. a. proficient
    b. theoretical
    c. efficient

19. a. bored
    b. enterprising
    c. active

20. a. involved
    b. vigorous
    c. docile

Turn to pages 60 to 65 for the answers.

Speed Test 2

→ Time allowed: 20 minutes maximum to complete tests 1 to 5.

EXERCISE 1

3	4	5	1	7	6
2	8	4	9	3	1
7	5	6	2	8	4
9	2	7	8	3	6
2	6	1	8	4	7
3	5	2	7	4	9

4	7	3	2	9	7
6	8	3	4	8	1
9	1	5	8	2	6
4	8	3	7	7	3
5	9	7	2	6	2
8	1	3	7	4	5

Find a string of four numbers in the left-hand grid that also occur in the same order, backward, forward, up, down, or diagonally in the right-hand grid.

EXERCISE 4

Y	S	A	E	P
P	E	P	I	L
I	O	L	O	A
H	C	A	L	M
S	S	N	A	P

N	M	Y	H	T
H	E	A	A	L
S	P	I	L	S
I	O	I	E	O
F	H	E	A	R

Which four-letter word that appears in the left-hand grid also appears in the right-hand grid? The words can appear backward, forward, up, down, or diagonally in either grid.

EXERCISE 2

Which number is two squares away from itself less two, three places away from itself divided by two, one place away from itself plus two, and three places away from itself multiplied by two?

12	16	9	3	13
10	4	24	19	7
11	14	26	1	17
12	2	8	6	5
22	20	15	18	40

EXERCISE 5

Which two numbers appear in the incorrect order in the grid to the right?

7	4	2	9	6	8
9	3	8	7	3	2
2	6	6	4	4	9
8	8	3	3	7	7
6	2	4	8	9	4
3	9	7	6	2	3

EXERCISE 3

Multiply the second lowest odd number in the right-hand grid by the second highest even number in the left-hand grid.

4	27	38	16
45	10	9	2
18	1	14	61
15	20	7	6

8	13	6	9
1	21	5	10
17	2	14	27
11	16	15	7

Turn to pages 60 to 65 for the answers.

Psychometric Test 4

Choose the one answer that you think is most applicable to yourself.

→ Time allowed: 15 minutes maximum to complete questions 1 to 15.

15 mins.

1. **When you are watching the news on television, which of the following most interests you?**
   a. politics
   b. items of local interest
   c. global issues

2. **How often do you follow your intuition when making decisions, rather than thinking things through carefully?**
   a. often
   b. very rarely
   c. occasionally

3. **An unusual work of art, such as an unmade bed, wins a major abstract art prize. What is your reaction?**
   a. interested
   b. appalled
   c. amused

4. **Assuming you have never written a novel, what is most likely to be your reaction if asked, "Do you think you could ever write a novel?"**
   a. I would like to, but am not sure if I could
   b. I feel I could, if I set my mind to it
   c. maybe, but I doubt if I would ever find the time

5. **To be described as which of the following would please you the most?**
   a. someone who is highly successful in their chosen profession
   b. a kind and caring family person
   c. a talented and amusing, free-thinking individual

6. **If you won $1 million in a lottery, which of the following statements would appeal to you the most?**
   a. I will invest for the long-term security of myself and my family
   b. I am now free to do all the things I always wanted to do
   c. I can now build up my own successful business

7. **The thought of which one of the following terrifies you the most?**
   a. being a total failure in life
   b. going off to fight in a war
   c. being locked up in prison for five years

8. **Of the following three school subjects, which one was your favorite?**
   a. sports
   b. art
   c. mathematics

9. **Which of these presents would please you most?**
   a. Yamaha portable keyboard
   b. two-week holiday in Disneyland
   c. laptop computer

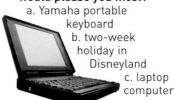

10. **When did you last take up a new hobby?**
    a. more than five years ago
    b. between two and five years ago
    c. less than two years ago

11. **What do you see as the main benefit of retirement?**
    a. freedom from the drudgery of going to work each day
    b. having more time to spend with your family and friends
    c. the time for new activities

12. **What do you see as the main drawback to retirement?**
    a. apart from the fact that I would be getting older, there are no big drawbacks
    b. I will miss my job and my colleagues
    c. I may be bored and have difficulty filling my time

13. **Do you believe in ESP (extrasensory perception)?**
    a. maybe
    b. yes
    c. no

14. **What is the best use of knowledge?**
    a. it broadens my horizons and enables me to obtain even more knowledge
    b. it enables a person to obtain qualifications and be successful in life
    c. it keeps the mind constantly active

15. **Which of the following do you think is the most important?**
    a. good common sense
    b. ambition
    c. a lively imagination

Turn to pages 60 to 65 for the answers.

Agility of Mind & Verbal Analogy Tests

→ Time allowed: 3 minutes maximum to complete questions 1 and 2.

AGILITY OF MIND TEST

1. Arrange the odd digits in ascending order, followed by the even digits in descending order.

For example:
3842197 = 1379842

a. 256934
b. 879461
c. 529648
d. 7419825
e. 3497216

2. Arrange the following letters in alphabetical order from the end of the alphabet to the beginning.

For example:
OCRFTI = TROIFC

1. TKVXBR
2. SWJGUCYQ
3. KZMCWHET
4. NLAVPFRD
5. KJBXSHVO

EXAMPLE

Choose the word that will correctly complete the analogy. For example:

Thursday is to Monday as July is to: May, April, February, March, September   Answer: April

→ Time allowed: 20 minutes maximum to complete questions 1 to 19.

VERBAL ANALOGY TEST

1. square is to cube as circle is to:
cone, sphere, octahedron, cylinder, circumference

2. pampas is to grassland as oxbow is to:
valley, lake, sea, hill, reef

3. chain is to saw as hand is to:
craft, hammer, press, tool

4. melanite is to black as alexandrite is to:
blue, green, white, red, purple

5. door is to threshold as roof is to:
tile, ridge, gable, eave, apron

6. pomace is to pulp as compote is to:
juice, peel, dry, stew, flesh

7. helmet is to protection as tiara is to:
royalty, adornment, crown, identification, devotion

8. perilune is to moon as perihelion is to:
earth, sun, planet, orbit, comet

9. hand is to clock as gnomon is to:
egg timer, sundial, calendar, metronome, clepsydra

10. garments is to tailor as candles is to:
fletcher, chandler, turner, cooper, spencer

11. new is to old as neo is to:
oleo, paleo, helico, tauto, eco

12. Nepal is to Hungary as Gurkha is to:
fusilier, cossack, chasseur, hussar, hessian

13. cultured is to charming as highbrow is to:
cosmopolitan, precocious, cultivated, stylish, urbane

14. songwriter is to songbird as lyricist is to:
passerine, music, paean, chorus, emollient

15. shock is to outburst as trauma is to:
hysteria, obsession, neurosis, paroxysm

16. snide is to coy as snigger is to:
simper, titter, guffaw, smile

17. room is to cup as cenacle is to:
platter, goblet, grail, glass

18. Kelvin is to Newton as temperature is to:
energy, power, heat, force

19. abet is to connive as facilitate is to
avail, invoke, intercede, succor, expedite

Turn to pages 60 to 65 for the answers.

Psychometric Test 5

Choose the one answer that you think is most applicable to yourself.

→ Time allowed: 15 minutes maximum to complete questions 1 to 15.

1. **If you had an unlimited amount of money, how often would you redecorate your living room?**
   a. twice every five years
   b. more than twice every five years
   c. once every five years

2. **How often do you put your thoughts down on paper?**
   a. quite often
   b. rarely
   c. occasionally

3. **Of the following, which is your favorite type of puzzle?**
   a. word searches
   b. jigsaw puzzles
   c. cryptic crosswords

4. **Which one of the following statements most accurately represents your views on graffiti?**
   a. in many ways it is a sad reflection of our times
   b. it is a great nuisance and perpetrators should be prosecuted
   c. it can be unsightly, but some of it can be artistic

5. **Assuming you have had an urge at some time to try a new creative hobby, such as painting or playing the piano, what have you done about it?**
   a. I did try, but didn't keep at it
   b. I tried it and it became one of my hobbies
   c. never got around to it

6. **Which one of the following statements can you identify with the most?**
   a. rules and regulations annoy me
   b. I like to create my own rules
   c. I am happy to live by already established rules

7. **Which one of these types of music do you enjoy most?**
   a. jazz
   b. pop
   c. opera

8. **Which one of the following statements appeals to you most?**
   a. I have a wonderful family life
   b. I do things because I want to do them
   c. I am a success in my chosen career

9. **With which of these sports personalities' attitudes do you most identify?**
   a. Tiger Woods
   b. Lance Armstrong
   c. John McEnroe

10. **Which of the following most accurately describes your attitude toward change?**
    a. change, like death and taxes, is inevitable
    b. things very rarely change for the better
    c. we should have a continuous commitment toward changing things for the better

11. **Which of these words appeals to you most?**
    a. serious
    b. curious
    c. organized

12. **Which of these best describes you?**
    a. methodical
    b. sensible
    c. setting yourself high standards

13. **Which of these best describes your attitude toward failure?**
    a. it can be depressing
    b. give up and try something new
    c. try, try, and try again

14. **How often do you daydream?**
    a. often
    b. sometimes
    c. rarely

15. **Which of the following would you choose for a day out?**
    a. giant ferris wheel
    b. labyrinth
    c. wax museum

Turn to pages 60 to 65 for the answers.

59

Answers to Section One

SPATIAL AWARENESS TEST
(test on pages 30 to 32)

1. **C**
2. **A & C**
3. **C & D**
4. **E**
5. **C**
6. **B**
7. **B**
8. **A & E**
9. **B & C**

Score 3 points for each correct answer, and 1 point for just one correct answer in questions 2, 3, 8, and 9.

Enter your score against *Spatial Awareness Test* in Column B of Appendix 1 on page 69.

MISSING PIECE TEST
(test on pages 33 to 35)

1. **D.** Each horizontal and vertical line contains two green stars, one pink circle, and one purple circle.
2. **A.** Each horizontal and vertical line contains one each of the four different squares.
3. **D.** Start at the top left-hand corner and work along the top line, then back along the next line, repeating the sequence orange, blue, green. Finish at the bottom left-hand corner square.
4. **B.** Looking across each line and down each column, a line is added to the figure in alternate squares.
5. **D.** Each horizontal line contains the symbol in each one of four positions.
6. **C.** Each of the symbols appears in each vertical, horizontal, and corner-to-corner line.
7. **D.** Each horizontal line contains one square that contains all the symbols in the other three squares.
8. **C.** Each horizontal and vertical line contains one each of the four different squares.
9. **C.** Each horizontal and vertical line contains three white circles and three black circles.
10. **B.** Alternate squares have their colored portion in opposite quarters.

Score 2 points for each correct answer.
Enter your score against *Missing Piece Test* in Column B of Appendix 1 on page 69.

PSYCHOMETRIC TEST 1
(test on page 36)

Award yourself the following points for each answer:

1. **a.** 1 point	**b.** 2 points	**c.** 0 points
2. **a.** 0 points	**b.** 1 point	**c.** 2 points
3. **a.** 0 points	**b.** 1 point	**c.** 2 points
4. **a.** 2 points	**b.** 0 points	**c.** 1 point
5. **a.** 2 points	**b.** 1 point	**c.** 0 points
6. **a.** 2 points	**b.** 0 points	**c.** 1 point
7. **a.** 2 points	**b.** 1 point	**c.** 0 points
8. **a.** 1 point	**b.** 0 points	**c.** 2 points
9. **a.** 2 points	**b.** 1 point	**c.** 0 points
10. **a.** 0 points	**b.** 2 points	**c.** 1 point
11. **a.** 1 points	**b.** 2 points	**c.** 0 points
12. **a.** 1 point	**b.** 0 points	**c.** 2 points
13. **a.** 2 points	**b.** 0 points	**c.** 1 point
14. **a.** 2 points	**b.** 1 point	**c.** 0 points
15. **a.** 1 point	**b.** 2 points	**c.** 0 points

Enter your score against *Psychometric Test 1* in Column A of Appendix 1 on page 69.

PSYCHOMETRIC TEST 2
(test on page 43)

Scoring:
Award yourself the following points for each answer:

0 points for every answer **a**
1 point for every answer **b**
2 points for every answer **c**

Enter your score against *Psychometric Test 2* in Column A of Appendix 1 on page 69.

NUMERICAL TEST
(test on page 37)

a. **127.** Add the sum of the digits of the previous number.

b. **10.** There are two sequences running alternately: 1, 4, 7, 10 and 2, 3, 4.

c. **212.** They are the odd numbers 1, 3, 5, and so on, divided into groups of three.

d. **13.** Add 1, 2, 3 and repeat.

e. **87.** Deduct 1, 2.5, 4, 5.5, and so on. Increase by 1.5 each time.

f. **33.75.** Two alternate sequences: the first add 1.5 and the second multiply by 1.5.

g. **63.** Multiply by 2 and add 1 each time.

h. **32.** Add the two previous numbers to obtain the third number.

i. **−1.** Add 1, subtract 2, and so on.

j. **77.** Two alternate sequences: add 6 and subtract 6.

k. **6.5.** Divide by 3, then add 3, multiply by 3, subtract 3, and repeat.

l. **14.641.** Multiply by 1.1 each time.

m. **0.** The sequence runs subtract 9, add 1, subtract 8, add 2, subtract 7, add 3.

n. **1.4.** Divide by 5, add 5, and so on.

o. **51.** Multiply by 3, add 2, and repeat.

p. **62.5.** Divide by 2 what you are deducting each time: 20, 10, 5, 2.5.

q. **71.** Add square numbers each time: 1, 4, 9, 16, 25.

r. **108.** Add last digit of previous number each time.

s. **187.5.** Multiply by 2.5 each time.

Score 1 point for each correct answer.
Enter your score against *Numerical Test 1* in Column A of Appendix 1 on page 69.

VISUAL ANALOGY TEST
(test on pages 38 to 40)

1. **B.** The color at the bottom goes in the middle, the color in the middle goes to the top, and the color at the top goes to the bottom.

2. **A.** Curved red lines turn to purple straight lines, and vice versa.

3. **D.** The color on the immediate right goes to the top of the pyramid, and the remaining two colors form the base of the pyramid but change places.

4. **A.** The number of triangles reduces from two to one and the number of squares increases from one to two. The triangle then goes between two squares above and below it.

5. **D.** The straight-sided figure rotates 90 degrees clockwise, and in the two figures the colors change places.

6. **E.** The colored portions change places according to their original position as in the first analogy.

7. **B.** The four figures originally in the center form one large figure. The large figure becomes four smaller figures and goes inside the new large figure.

8. **C.** The bottom and top two lines move up and down respectively, the line previously at the bottom rests on the two lines previously at the top.

9. **E.** The yellow circles become green triangles, and vice versa.

Score two points for each correct answer.
Enter your score against *Visual Analogy Test* in Column B of Appendix A on page 69.

SPEED TEST 1
(test on pages 41 to 42)

1. True	11. True	21. True
2. False	12. False	22. True
3. True	13. True	23. True
4. False	14. True	24. False
5. False	15. True	25. False
6. False	16. False	26. True
7. True	17. True	27. True
8. False	18. False	28. True
9. True	19. True	29. True
10. True	20. True	

Score 1 point for each correct answer.
Enter your score against *Speed Test 1* in Column A of Appendix 1 on page 69.

Answers to Section One

VISUAL CLASSIFICATION TEST
(test on pages 44 to 45)

••

1. **E.** The dot appears in the square and two circles.
2. **B.** Triangle in circle, circle in triangle, and black dot in square.
3. **A.** One dot appears in two circles and the other dot appears in a circle and a triangle.
4. **A.** A circle appears in the largest segment and a black dot in the smallest segment.
5. **C.** One dot appears in a square and circle, and the other dot appears in a triangle and square.
6. **D.** One dot appears in a triangle and the other appears in a five-sided figure.
7. **E.** One dot appears in three circles and the other dot appears in one circle and the triangle.
8. **C.** The circle string is black/white/black/white/white.
9. **D.** It has both vertical and lateral symmetry. In other words, the top half is a mirror image of the bottom, and the right side is a mirror image of the left.

Score 2 points for each correct answer.
Enter your score against *Classification Visual Test* in Column B of Appendix 1 on page 69.

LITTLE PUZZLES TEST
(test on pages 46 to 47)

••

1. **32.** Each number represents its position in the grid, so 32 is on row 3, column 2.
2. **3145.**

A	B	C	D		A	B	C	D
1	5	3	4		3	1	4	5

3. None of them repeat a letter.
4. It is a traffic island.
5. **11.** Each group of three triangular blocks A, B, C totals 21.

6. Veracruz. Each place starts with the middle two letters of the place previously visited.
7. **8.** 3+5 = 8. The numbers in the blue sections to each side of the middle drawing make up the number 8.

8. **6.** Each ring contains the numbers 0 to 9 only once.
9. Start at the top left-hand corner square and work clockwise round the perimeter, spiraling into the center, repeating the same four digits.

10. **6492875.** The last four digits of the previous number followed by the first three.
11. **27.6.** Add on the sum of digits of the previous number as a decimal.
12. **E.** They are the first letters in the alphabet that can be written with 1, 2, 3, and 4 lines respectively.
13. Start at A and follow the route shown above, omitting alternate letters of the alphabet.

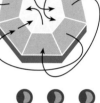

14. They are the numbers 7, 6, 5, 4 written digitally and on their side.

Score 2 points for each correct answer.
Enter your score against *Little Puzzles Test* in Column B of Appendix 1 on page 69.

VERBAL TEST
(test on page 48)

••

1. irrational, wise
2. stereotype, pigeonhole
3. incredible
4. nostrum, panacea
5. simple
6. myriad
7. erroneous, faultless
8. polluted
9. verve, zeal
10. pragmatic
11. confirm
12. agog, impatient
13. condone, censure
14. viscid
15. piercing, mellifluous
16. ingenuous
17. petulant
18. redundant
19. rosy, wan

Score 1 point for each correct answer.
Enter your score against *Verbal Test 1* in column A of Appendix 1 on page 69.

VISUAL SEQUENCE TEST
(test on page 49)

1. **E.** There are two alternate sequences: red triangle with blue dot and blue triangle with red dot. In each sequence the dot moves to a different corner counterclockwise.
2. **D.** The red dot moves 90 degrees counterclockwise at each stage, and the turquoise dot moves 180 degrees.
3. **B.** A new quarter segment is added at each stage, then the two colors in this segment change places at each stage.
4. **E.** It is a mirror image of the second figure, as the third figure is of the first.
5. **A.** A new dot is added each time as if at the corners of a square. Each dot added is a new color at the top left-hand corner. The previous dots move one corner counterclockwise to make room for it.
6. **F.** The dots are being repeated in the sequence green/yellow/red/purple/brown.
7. **C.** The total number of sides in each set of two figures increases by two each time.
8. **B.** Each figure moves from top to bottom, one place at each stage. While it is moving the figure changes from red to yellow.
9. **D.** Each arm moves 90 degrees clockwise at each stage.

Score 2 points for each correct answer.
Enter your score against *Visual Sequence Test* in Column A of Appendix 1 on page 69.

VISUAL SHAPES TEST
(test on pages 51 to 52)

1.

2.

3.

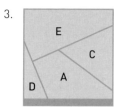

4.

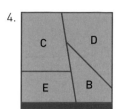

5.

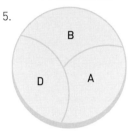

6.

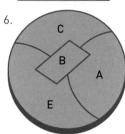

7.

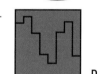 **D**

8.

D

9.

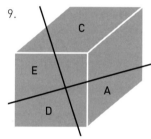

Score 2 points for each correct answer. Enter your score against *Shapes Visual Test* in Column B of Appendix 1 on page 69.

WORD DEFINITION TEST
(test on page 53)

1. **d.** a familiar old proverb
2. **a.** printed pamphlets
3. **c.** doubt
4. **c.** jaunty
5. **b.** unable to blend
6. **a.** conjectural
7. **d.** superficial
8. **c.** aggressive
9. **b.** ghost
10. **d.** ring or eyelet
11. **b.** the big toe
12. **a.** to recommend
13. **d.** truthfulness
14. **b.** tightly packed
15. **a.** elaborate
16. **d.** person of brilliant achievement
17. **b.** twilight
18. **c.** sticky
19. **a.** to encourage or provoke

Score 1 point for each correct answer.
Enter your score against *Word Definition Test* in Column A of Appendix 1 on page 69.

Answers to Section One

Visual Creativity Test
(test on page 54)

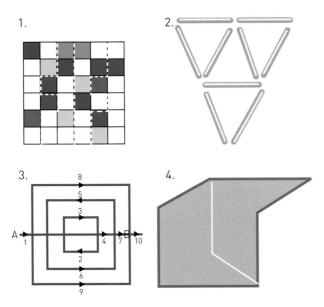

1.

2.

3.

4.

5. Circle 3

6. Each color appears eight times. Because the same color is never adjoining, each of the eight cubes must contain each of the three colors only once on its external sides.

7. Yellow

8.

9. **E.** All the others are divided into four equal segments.

Score 3 points for each correct answer.
Enter your score against *Visual Creativity Test* in Column B of Appendix 1 on page 69.

Psychometric Test 3
(test on page 55)

Award yourself the following points for each answer:

1.	**a.** 0 points	**b.** 1 point	**c.** 2 points
2.	**a.** 1 point	**b.** 0 points	**c.** 2 points
3.	**a.** 2 points	**b.** 1 point	**c.** 0 points
4.	**a.** 0 points	**b.** 2 points	**c.** 1 point
5.	**a.** 2 points	**b.** 1 point	**c.** 0 points
6.	**a.** 0 points	**b.** 1 points	**c.** 2 points
7.	**a.** 0 points	**b.** 1 point	**c.** 2 points
8.	**a.** 2 points	**b.** 1 point	**c.** 0 points
9.	**a.** 0 points	**b.** 2 points	**c.** 1 point
10.	**a.** 2 points	**b.** 1 point	**c.** 0 points
11.	**a.** 1 point	**b.** 2 points	**c.** 0 points
12.	**a.** 1 point	**b.** 0 points	**c.** 2 points
13.	**a.** 0 points	**b.** 1 point	**c.** 2 points
14.	**a.** 1 point	**b.** 2 points	**c.** 0 points
15.	**a.** 1 point	**b.** 0 points	**c.** 2 points
16.	**a.** 0 points	**b.** 1 point	**c.** 2 points
17.	**a.** 1 point	**b.** 0 points	**c.** 2 points
18.	**a.** 2 points	**b.** 0 points	**c.** 1 point
19.	**a.** 2 points	**b.** 0 points	**c.** 1 point
20.	**a.** 1 point	**b.** 2 points	**c.** 0 points

Enter your score against *Psychometric Test 3* in column A of Appendix 1 on page 69.

Speed Test 2
(test on page 56)

1. 2783

2. 6

3. 100 (20 x 5)

4. HOPE

5.

Start at the bottom left-hand corner and travel up the first column, then back down the second, repeating the numbers 3682974.

The 6 and 8 should be transposed as shown in the diagram to the left.

Score 4 points for each correct answer. Enter your score against *Speed Test 2* in column A of Appendix 1 on page 69.

AGILITY OF MIND TEST
(test on page 58)

	1.	2.	
	a. 359642	1. XVTRKB	Score 1 point for each correct answer. Enter your score against *Agility of Mind Test* in Column A of Appendix 1 on page 69.
	b. 179864	2. YWUSQJGC	
	c. 598642	3. ZWTMKHEC	
	d. 1579842	4. VRPNLFDA	
	e. 1379642	5. XVSOKJHB	

VERBAL ANALOGY TEST
(test on page 58)

1. sphere	10. chandler	19. expedite
2. lake	11. paleo	Score 1 point for each correct answer. Enter your score against *Verbal Analogy Test* in Column B of Appendix 1 on page 69.
3. press	12. hussar	
4. green	13. urbane	
5. eave	14. passerine	
6. stew	15. paroxysm	
7. adornment	16. simper	
8. sun	17. grail	
9. sundial	18. force	

PSYCHOMETRIC TEST 4
(test on page 57)

Award yourself the following points for each answer:

1. **a.** 1 point **b.** 0 points **c.** 2 points
2. **a.** 2 points **b.** 0 points **c.** 1 point
3. **a.** 2 points **b.** 0 points **c.** 1 point
4. **a.** 1 point **b.** 2 points **c.** 0 points
5. **a.** 0 points **b.** 1 point **c.** 2 points
6. **a.** 1 point **b.** 2 points **c.** 0 points
7. **a.** 0 points **b.** 2 points **c.** 1 point
8. **a.** 1 point **b.** 2 points **c.** 0 points
9. **a.** 2 points **b.** 1 point **c.** 0 points
10. **a.** 0 points **b.** 1 point **c.** 2 points
11. **a.** 1 point **b.** 0 points **c.** 2 points
12. **a.** 2 points **b.** 0 points **c.** 1 point
13. **a.** 1 point **b.** 2 points **c.** 0 points
14. **a.** 2 points **b.** 0 points **c.** 1 point
15. **a.** 1 point **b.** 0 points **c.** 2 points

Enter your score against *Psychometric Test 4* in Column B of Appendix 1 on page 69.

PSYCHOMETRIC TEST 5
(test on page 59)

Award yourself the following points for each answer:

1. **a.** 1 point **b.** 2 points **c.** 0 points
2. **a.** 2 points **b.** 0 points **c.** 1 point
3. **a.** 2 points **b.** 1 point **c.** 0 points
4. **a.** 1 point **b.** 0 points **c.** 2 points
5. **a.** 1 point **b.** 2 points **c.** 0 points
6. **a.** 1 point **b.** 2 points **c.** 0 points
7. **a.** 2 points **b.** 1 point **c.** 0 points
8. **a.** 1 point **b.** 2 points **c.** 0 points
9. **a.** 1 point **b.** 0 points **c.** 2 points
10. **a.** 1 point **b.** 0 points **c.** 2 points
11. **a.** 0 points **b.** 2 points **c.** 1 point
12. **a.** 1 point **b.** 0 points **c.** 2 points
13. **a.** 0 points **b.** 1 point **c.** 2 points
14. **a.** 2 points **b.** 1 point **c.** 0 points
15. **a.** 1 point **b.** 2 points **c.** 0 points

Enter your score against *Psychometric Test 5* in Column B of Appendix 1 on page 69.

Analysis and Assessment

The analysis of the scores obtained on all the tests in section one reveals whether you have a left-side or right-side bias and by how much. Your weighting may, on the other hand, appear in the balanced middle band.

There is no right or wrong score on this graph. If you are right-side biased, you are likely to have a strong intuitive and creative nature; if you are left-side biased, you are likely to be analytical and logical, with good numerical and verbal skills.

If your score shows you are within the balanced-brain band, this is no cause for complacency. One problem with hemispheric balance is that you may feel conflict between what you feel and what you think. The advantage is that you can perceive the big picture and the essential details simultaneously, and that you are likely to possess the verbal skills to translate your intuition into an understandable form.

Just because you are heavily weighted to the right side does not mean that you are strong in every predominant right-brain skill; no one is entirely left- or right-brained. You need to analyze your individual scores carefully in each of the tests by referring to the scoring chart on page 69.

Scores registered in column A are for left-hand discipline tests, and scores in column B are for right-hand discipline tests. The higher your score on each of these tests, the stronger you are at that particular discipline. For example, while your score might indicate that you have a heavy bias toward left-side dominance, it may still be that you scored badly on

---

To summarize:

Left hemisphere

- Analytic
- Verbal
- Rational
- Logical
- Sequential, linear
- Time-oriented
- Tendency toward science and mathematics
- Frank, direct
- True, unfanciful
- Sensible
- Forceful

→ **The left hemisphere** absorbs, analyzes, and processes information logically and sequentially.

the numerical tests in section one and, therefore, you need to work at that particular skill.

BALANCING YOUR BRAIN

In general, the left and right hemispheres of the brain handle information in different ways, and we tend to address information using our dominant side. The learning and thinking process is enhanced when both sides of the brain work in a balanced manner. This means that you need to strengthen your less dominant hemisphere, which is the purpose of the tests in section two.

As already explained, for most people the left side of the brain is analytical and functions in a sequential and rational fashion, and is the side that controls language, academic studies, and rationality. On the other hand, the right side is creative and intuitive and leads to the birth of ideas for works of art and

→ **THE CREATIVE RIGHT HEMISPHERE** needs to diversify continually, to explore new avenues, and to uncover artistic talents.

RIGHT HEMISPHERE

- Intuitive, imaginative
- Spatial
- Synthetic
- Creative, artistic
- Simultaneous, holistic
- Timeless, spiritual
- Tendency toward music, art, dance
- Flexible
- Playful, fanciful
- Complex

← **COMPOSERS, LIKE** writers, find many of their creative ideas from within the subconscious right hemisphere.

Analysis and Assessment

music. And this is where the interface between the two halves of the brain becomes so important. In order for the subconscious of your right hemisphere to function, it needs the fuel, or data, that has been fed into, collected, and processed by the left hemisphere.

The danger is overburdening the left hemisphere with too much data, and too quickly, to the extent that the creative side of your brain is unable to function to its full potential. By contrast, a lack of data fed into the left hemisphere could result in the creative side, or right hemisphere, drying up. It is, therefore, desirable to strike the right balance between the two hemispheres in order for the brain to work to its full potential.

↓ **SECTION ONE TESTS** will determine which side of your brain is dominant.

↑ **SECTION TWO PUZZLES** will exercise both the left and right hemispheres and ultimately balance the two sides for maximum performance.

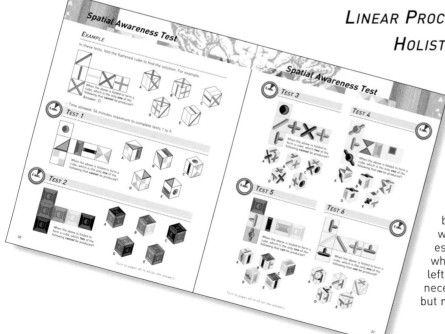

LINEAR PROCESSING (LEFT) VERSUS HOLISTIC PROCESSING (RIGHT)

By processing information in a linear manner, the left side of the brain processes from part to whole. The right brain processes holistically, by seeing the big picture and not the component details. When attending a lecture, for example, right-brained people may be at a disadvantage unless they have been given an overview of the whole concept first, because they essentially need to know exactly what they are doing and why. The left-brain student may not find it necessary to look this far ahead, but may find it helpful to do so.

Number	Test description	Column A (Left-brain)	Column B (Right-brain)
1	Spatial Awareness		
2	Missing Piece		
3	Psychometric 1		
4	Numerical		
5	Visual Analogy		
6	Speed 1		
7	Psychometric 2		
8	Visual Classification		
9	Little Puzzles		
10	Verbal		
11	Visual Sequence		
12	Visual Shapes		
13	Word Definition		
14	Visual Creativity		
15	Psychometric 3		
16	Speed 2		
17	Psychometric 4		
18	Agility of Mind		
19	Verbal Analogy		
20	Psychometric 5		

Add your score in each column and calculate the difference between the two columns.
A plus score in the left-hand column A indicates a left-brain bias.
A plus score in the right-hand column B shows a right-brain bias.

Read your score on the graph below to find the extent of your right- or left-brain bias.

For example:
+70 in column A indicates high left-brain bias;
+20 in column B indicates slight right-brain bias, but is within the balanced-brain reading.

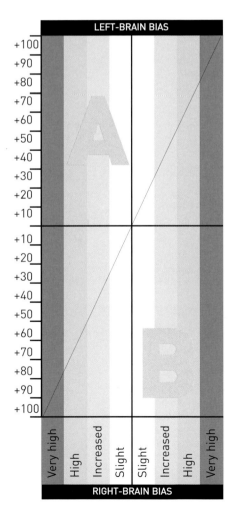

Analysis and Assessment

Logic (Left) versus Intuition (Right)

In addition to its linear and sequential approach, the left brain processes in a logical manner—it uses information piece by piece to work out, for example, a mathematical problem. The right side of the brain tends to use intuition—a right-brain person may know the answer to a question, but is not quite sure how the answer was reached.

When writing, the left-brain person pays careful attention to spelling and punctuation, while the right-brain person pays attention to the overall meaning, and whether what has been written feels right.

← **LIKE A COMPUTER,** the left hemisphere processes data sequentially and logically.

→ **ATTENTION TO DETAIL** makes many left-brain people fastidious note takers in a classroom or meeting.

Sequential Processing (Left) versus Random Processing (Right)

The left brain tends to process things in sequence, while the approach of the right-brain person is random. The left-brain person is likely to be a fine planner or accountant. Spelling is also likely to be a strong point of the left-brain person because it involves sequencing.

In order to strengthen their sequencing skills, right-brain people should attempt to make lists and schedules, which will discipline them to complete tasks more efficiently, and without flitting from one assignment to another.

In the classroom, right-brain students may feel at a disadvantage because this is where left-brain strategies are most often used. However, these students can develop enough flexibility to adapt material to the right side of the brain. In the same way, predominantly left-brain readers should realize that it is possible and desirable to use both sides of the brain and employ some right-brain strategies.

Now work through the exercises in section two at your own pace, concentrating first on the exercises that are designed to strengthen the side of your brain that is less dominant. However, in order to attain peak condition and guarantee a truly balanced approach, it is recommended that you complete all the exercises in this section, both left and right.

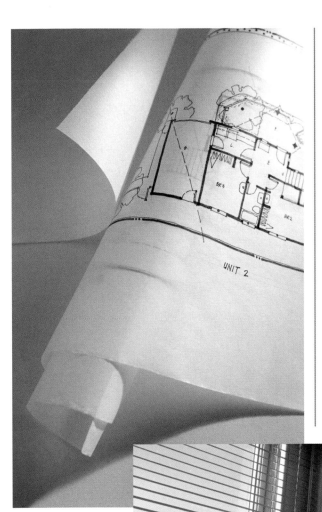

↑ **LEADERSHIP QUALITIES,** like Napoleon's, must be combined with brilliant organizational skills in order to turn ambitious dreams into reality.

↑→**AN ARCHITECT** needs to balance creativity with logic and detail to turn a concept into a workable reality.

WORKOUT FOR A
BALANCED
BRAIN

Now that you know which side

of your brain is dominant,

readjust the balance by exercising the side

that you use less on a daily basis.

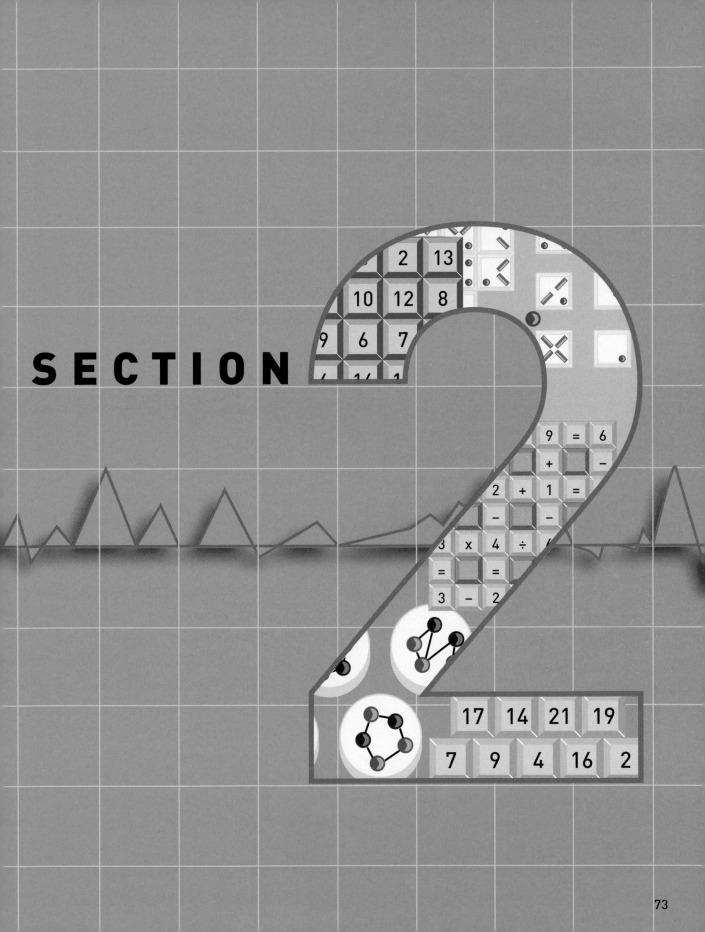

SECTION 2

EXAMPLE

Insert the numbers listed in each puzzle into the circles so that—for any particular circle— the sum of the numbers in the circles connected to it equals the value corresponding to that circled number in the list.

Answer: 1 = 14 (4+7+3)
4 = 8 (7+1)
7 = 5 (4+1)
3 = 1

EXERCISE 1

→ Insert the numbers 1–7

1 = 11
2 = 9
3 = 19
4 = 10
5 = 3
6 = 7
7 = 3

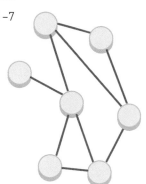

EXERCISE 2

→ Insert the numbers 1–7

1 = 12
2 = 14
3 = 9
4 = 16
5 = 14
6 = 3
7 = 11

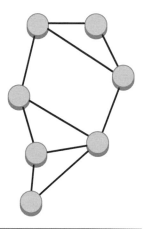

EXERCISE 3

→ Insert the numbers 1–8

1 = 32
2 = 1
3 = 14
4 = 16
5 = 8
6 = 4
7 = 10
8 = 8

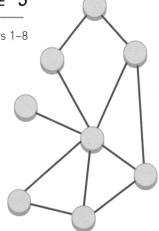

EXERCISE 4

→ Insert the numbers 1–9

1 = 16
2 = 27
3 = 23
4 = 18
5 = 2
6 = 10
7 = 16
8 = 5
9 = 5

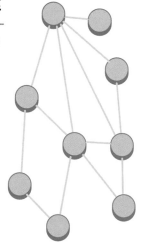

Continued on page 76

Exercises to improve the right brain

FOR THOSE WITH A LEFT-BRAIN BIAS

CREATIVE WRITING

→ Many people would like to write a novel, yet only a very small percentage make it past the initial idea. In order to perform any creative task, you will need to encourage your right side to start its creative juices flowing.

One way to do this is to lull the left brain into a degree of inactivity—or even bore it to sleep. This could be done on a long train or plane trip, when the right brain has the opportunity to become more creative because it has less opposition from the fact-cluttered dominant left side. Or make sure you are in a secluded place, with no distracting phone calls or conversations. Ideas crop up at night, too, before you sleep, and as you dream.

Keep a pad and pen with you, and make a note of all the ideas that come into your head. Don't evaluate them to decide if they are useful or if the spelling is correct. Just let go, and allow the words and ideas to flow. When you have written all you can,

read it through, then reread it; start expanding on various parts that spark your imagination and creativity.

Now reawaken your left brain to organize what you have put down on paper. Accept that it will be something of a mess. This is the easiest thing to sort out and this is what a left-brain dominant person is good at. The main thing is that you have created an original work.

NOW TRY THE FOLLOWING EXERCISE:
The two paintings below depict a number of characters. Choose either one or two of these characters and try to write 1,000 to 3,000 words on each of them by incorporating the techniques above.

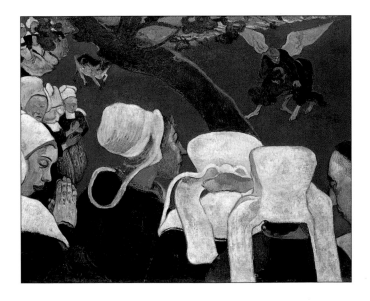

Continued from page 74

EXERCISE 5

→ Insert the numbers 1–7

1 = 17
2 = 11
3 = 1
4 = 2
5 = 14
6 = 7
7 = 6

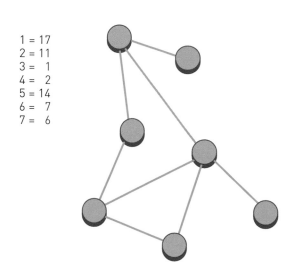

EXERCISE 6

→ Insert the numbers 1–7

1 = 11
2 = 22
3 = 5
4 = 8
5 = 9
6 = 2
7 = 3

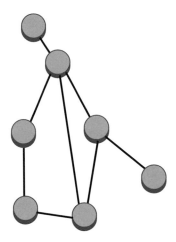

EXERCISE 7

→ Insert the numbers 1–7

1 = 14
2 = 16
3 = 14
4 = 2
5 = 10
6 = 4
7 = 10

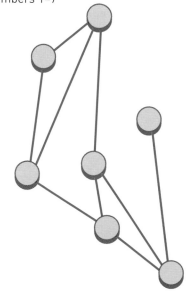

EXERCISE 8

→ Insert the numbers 1–7

1 = 10
2 = 5
3 = 12
4 = 14
5 = 13
6 = 11
7 = 15

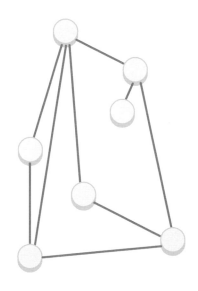

Continued on page 78

LATERAL NUMBER PUZZLES

EXAMPLE

These number puzzles are designed to exercise your powers of lateral thinking and creativity. None of these puzzles involves more than a very basic knowledge of mathematics. What they do involve, however, is an ability not to take things at face value, and to be able to think sideways. If you cannot solve these puzzles at the first attempt, return to them at a later stage and have another try.

For example:
What number is missing from the square?

Answer:
8. Each number is double the number of sides of the figure in which it is contained.

EXERCISE 1

→ What number should replace the question mark?

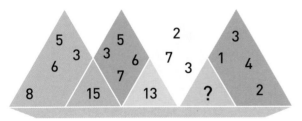

EXERCISE 2

→ What number should replace the question mark?

45
35   69
46   28
?   17
36

EXERCISE 3

→ What number should replace the question mark?

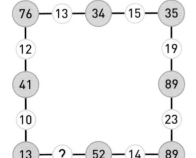

76 — 13 — 34 — 15 — 35
12   19
41   89
10   23
13 — ? — 52 — 14 — 89

EXERCISE 4

→ What number should replace the question mark?

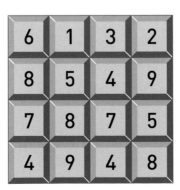

6	1	3	2
8	5	4	9
7	8	7	5
4	9	4	8

4	9	7	8
2	5	6	1
3	2	3	5
6	1	6	?

Continued on page 79

Turn to pages 144 to 155 for the answers.

MATHEMATICAL CALCULATION: CONNECTIONS

Continued from page 76

EXERCISE 9

→ Insert the numbers 1–8

1 = 20
2 = 15
3 = 11
4 = 8
5 = 11
6 = 10
7 = 22
8 = 14

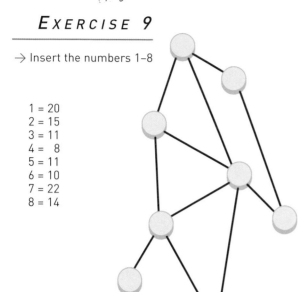

EXERCISE 10

→ Insert the numbers 1–8

1 = 17
2 = 5
3 = 13
4 = 14
5 = 18
6 = 10
7 = 8
8 = 5

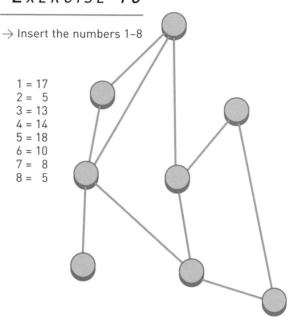

EXERCISE 11

→ Insert the numbers 1–9

1 = 4
2 = 15
3 = 26
4 = 19
5 = 8
6 = 5
7 = 24
8 = 16
9 = 12

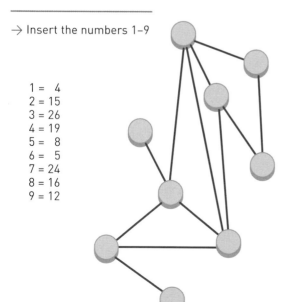

EXERCISE 12

→ Insert the numbers 1–9

1 = 18
2 = 26
3 = 8
4 = 3
5 = 15
6 = 25
7 = 11
8 = 10
9 = 20

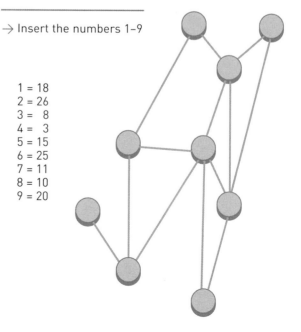

Continued from page 77

LATERAL NUMBER PUZZLES

EXERCISE 5

→ What number should replace the question mark?

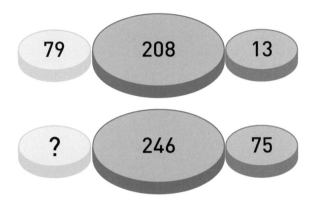

EXERCISE 6

→ What number should replace the question mark?

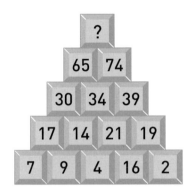

EXERCISE 7

→ What two numbers should replace the question marks?

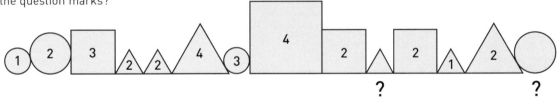

EXERCISE 8

→ What number should replace the question mark?

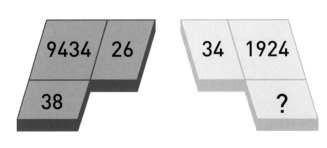

EXERCISE 9

→ What number should replace the question mark?

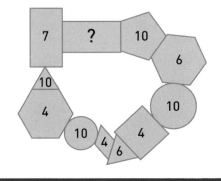

Continued on page 81

Turn to pages 144 to 155 for the answers.

Exercises to improve the left brain

DEFINITIONS TEST 1

→ Column A is a list of 10 definitions. Column B is a list of words that fit these definitions but in the wrong order. Put the correct word in column C to match its definition in column A.

A	B	C
female demon	hortatory	....
relating to animal fat	monition	....
fleshy part of plants	anodyne	....
small, delicate, and charming object	lamia	....
over-fussy attention to details	herbage	....
fine parchment	paladin	....
warning of danger	adipose	....
encouraging, urging on	vellum	....
heroic knight or hero	punctilio	....
bland or uncontroversial	bijou	....

DEFINITIONS TEST 2

→ Column A is a list of 10 definitions. Column B is a list of words that fit these definitions but in the wrong order. Put the correct word in column C to match its definition in column A.

A	B	C
to turn aside	immanent	....
military store	persecute	....
abiding in, inherent	ordinance	....
show off	imminent	....
subject to persistent ill treatment	ordnance	....
bring legal proceedings against	flout	....
decree	distract	....
to take away from	flaunt	....
show contempt for	prosecute	....
impending, close at hand	detract	....

Turn to pages 144 to 155 for the answers.   Continued on page 82

Continued from page 79
LATERAL NUMBER PUZZLES

EXERCISE 10

→ What number should replace the question mark?

EXERCISE 11

→ What number should replace the question mark?

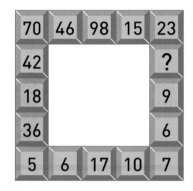

EXERCISE 12

→ What number should replace the question mark?

EXERCISE 13

→ What number should replace the question mark?

EXERCISE 14

→ What number should replace the question mark?

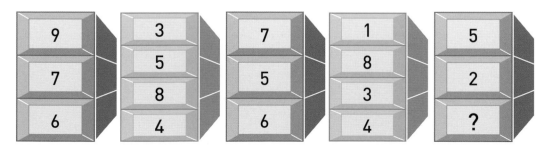

Continued on page 83

Exercises to improve the left brain
FOR THOSE WITH A RIGHT-BRAIN BIAS

DEFINITIONS TEST 3 *Continued from page 80*

→ Column A is a list of 10 definitions. Column B is a list of words that fit these definitions but in the wrong order. Put the correct word in column C to match its definition in column A.

A	B	C
cleverly contrived	loathe	....
standing still	judicial	....
in that place	stationery	....
belonging to them	loath	....
writing materials	ingenious	....
impartial	their	....
unwilling	ingenuous	....
prudent	stationary	....
naive, innocent	judicious	....
dislike intensely	there	....

DEFINITIONS TEST 4

→ Column A is a list of 10 definitions. Column B is a list of words that fit these definitions but in the wrong order. Put the correct word in column C to match its definition in column A.

A	B	C
supplies, especially military	principal	....
servile, ingratiating	misogamist	....
chief	precedence	....
one who hates women	misogynist	....
priority	materiel	....
substance of which something is made	obsequial	....
to do with funeral rights	principle	....
examples	material	....
one who hates marriage	precedents	....
fundamental truth	obsequious	....

Exercises to improve the right brain

FOR THOSE WITH A LEFT-BRAIN BIAS

Continued from page 81

LATERAL NUMBER PUZZLES

EXERCISE 15

→ What number should replace the question mark?

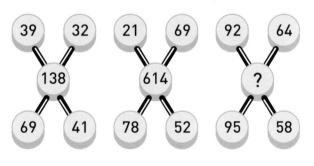

EXERCISE 16

→ What number should replace the question mark?

EXERCISE 17

→ What number should replace the question mark?

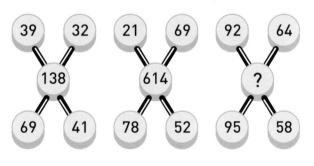

EXERCISE 18

→ What number should replace the question mark?

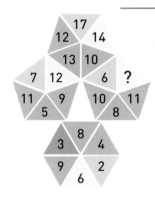

EXERCISE 19

→ What number should replace the question mark?

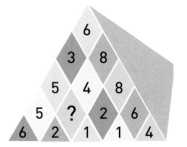

EXERCISE 20

→ What number should replace the question mark?

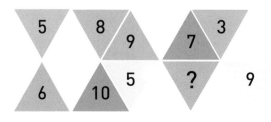

MATHEMATICAL CALCULATION

→ Many academic pursuits deal with symbols such as letters, words, and mathematical notations. The left-brain person tends to be comfortable with linguistic and mathematical endeavors because the left brain is more able to cope with processing the symbols.

A left-brain student is more likely to just memorize vocabulary words or mathematical formulas, while the right-brain person wants to see, feel, or touch the real object.

The mathematical puzzles here are designed to give the right-hemisphere person a hands-on experience of straightforward mathematical calculation. There is no trickery involved in solving these puzzles, just straightforward analysis and calculation.

1. My watch shows the time as 1:15 P.M.; one clock shows 1:25 P.M.; the stove clock shows 1:55 P.M.; the church clock strikes 2:00 P.M., and your watch shows 1:10 P.M. The radio announces the correct time as 1:30 P.M.
What is the average time—fast or slow—as shown by the above timepieces?

2. Jim has $3 more than Sid, but if Sid had three times more than he has now, he would have $12 more than the original combined amounts of money.
How much does Jim have?

3. A car travels 80 miles in the same time as another car traveling 20 mph faster covers 120 miles.
How long does the journey take?

4. If a car had increased its average speed for a 210-mile journey by 5 mph, the journey would have been completed in one hour less.
What was the original speed of the car for the journey?

5. I have four books standing on my desk between two bookends. If I decided to arrange the books in every possible different order, and it took two seconds to change the books each time, how long would it take me to arrange the books in every possible different combination?

6. Sid has $420 to spend. He spends three-fifths of the $420 on electrical goods, 0.45 of the remainder on clothes and writes out a check for $90 for a new watch.
What is his financial situation at the end of the day?

7. How many minutes before noon is it if 84 minutes ago it was three times as many minutes past 9 A.M.?

8. Tom, Dick, and Harry supply capital in a new business venture of $15,000, $30,000, and $55,000 respectively, and agree to share profits in proportion with the capital invested. Last year, $140,000 profits were available.
How much of the profit was allocated to each man?

9. Tony has a quarter again more than Sally, who has a quarter again more than Pat. Altogether they have 427.
How many do they each have?

10. I am now twice as old as my son. Twelve years ago I was five times as old as my son.
How much older was I than my son eight years ago?

Continued on page 86

Exercises to improve the right brain

FOR THOSE WITH A LEFT-BRAIN BIAS

LETTER AND WORD PUZZLES

EXAMPLE

All the letter and word puzzles in this exercise require a degree of lateral thinking. These exercises are not designed to test your knowledge of word meanings—a left-brain function—but to test your flexibility of thought instead. Several require analysis of the puzzle itself, a closer inspection for hidden patterns or meanings, and finally, an exploration of the puzzle beyond its visible boundaries in the search for a solution.

EXERCISE 1

→ Which word comes next in this sequence?

squirrel, squash, streak, house

Is it:

prove, print, charm, castle, or stair?

EXERCISE 2

→ What letter should replace the question mark?

EXERCISE 3

→ What do these phrases have in common?

sweet-talk, closed doors, coffee table, belle epoque

Continued on page 87   Turn to pages 144 to 155 for the answers.

11. **Jim is four times as old as Sally who is 4 years old.**
    How old will Jim be when he is twice as old as Sally?

12. **Three men are sitting down to a meal. John has seven loaves of bread. Henry has five loaves of bread. Charlie has no loaves but has $10.**
    If the loaves are shared equally, how much should Charlie give to each of the other two?

13. **Three men own a boat. Man A owns 42 percent, Man B owns 37 percent, and Man C owns 21 percent. An insurance bill is received for $300.**
    How much should each pay?

14. **Six horses are running a race.**
    In how many ways can they pass the post, assuming there are no dead heats?

15. **One man can drink a barrel of beer in 80 days. His wife can drink a barrel of beer in 200 days.**
    If they drink together at the same rate, how long will it take them to empty the barrel?

16. **In how many ways can a team of five be selected out of nine players?**

17. **Cyril has $50. Basil has 20 percent more than Cyril. George has 50 percent of the amount that Cyril and Basil have. Fred has 20 percent of the amount that Basil and George have.**
    How much have they together?

18. **Man A can build a wall in 4 hours. Man B can build a wall in 5 hours. Man C can build a wall in 10 hours.**
    If they all work together at the same rate, how long will it take to build the wall?

19. **Barbara was asked by a friend how old she was. Barbara said if I double my age and subtract one it would be the same as my uncle's age. If you reverse the digits of his age, you get my age. How old was Barbara?**

20. **How are the chances of scoring exactly 15 with three throws of the dice?**

21. **What are the chances of drawing four aces from a pack of 15 cards?**

22. **Bill hits the bull's-eye 80 times in 100 shots. Fred hits the bull's-eye 90 times in 100 shots.**
    What are the chances of the target being hit if each tries only once?

23. **All nine digits 1 to 9 inclusive can be used once each in the correct order using the + and − symbols only to equal 100.**
    There are 12 ways to do it. Can you find one?

24. **Find the sum of the numbers from 1 to 100 inclusive. There is a simple way.**

25. **In my lake I have 4,998 fish. The male fish have 111 spots each. The female fish have 37 spots each.**
    If I take out two-thirds of the male fish, how many spots are left in the lake?

26. **Insert the same number twice to make this calculation correct: 6 ÷ 8 = 27**

27. **Bill had $75 and three-quarters of what Alan had. Alan had $50 and half of what Bill had.**
    How much did each have?

28. **Insert the numbers 1 to 9 inclusive, once each, to arrive at an answer of 100.**

    $$(\quad - \quad) \ + \ + \ - \ - \ -$$

29. **Three men toss a coin. The winner is the first one to throw a tail. A goes first. B goes second, and C goes third.**
    What are the chances of each one winning?

30. **Substitute numbers for the letters to complete the sum.**

    ```
        W E A R Y
        L A W Y E R
      + R E A L L Y
        Y A W N E D
    ```

Continued from page 85

LETTER AND WORD PUZZLES

EXERCISE 4

→ What letter should replace the question mark?

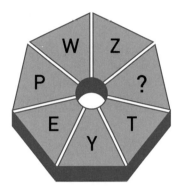

EXERCISE 5

→ Which three letters should go in the third oval?

EXERCISE 6

→ What letter should replace the question mark?

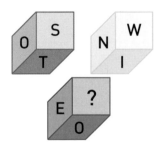

EXERCISE 7

→ Which two of these words do not belong in the group?

ago, elaborate, sat, rob,

sentenced, outplayed, net,

fantastic, lap, arrogance

EXERCISE 8

→ What letter should replace the question mark?

A, E, F, M, ?, W, Y

EXERCISE 9

→ What letter should replace the question mark?

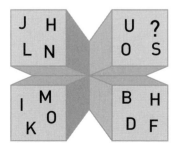

Turn to pages 144 to 155 for the answers.

Exercises to improve the left brain

FOR THOSE WITH A RIGHT-BRAIN BIAS

VERBAL INTELLIGENCE

EXAMPLE

Life today is all about communicating properly, and to do this effectively you need to increase your level of verbal intelligence and dexterity.

1. **Which word in brackets is closest in meaning to the word in capitals?**

   INVULNERABLE (susceptible, exposed, safe, confident, willful)

2. **Which word in brackets is the most opposite to the word in capitals?**

   POTENT (underdeveloped, ineffective, sad, unconditioned, deficient)

3. **Which two words are closest in meaning?**

   enthusiasm, fantasy, ballyhoo, delusion, style

4. **Which two words are closest in meaning?**

   absorb, jacket, wreathe, distort, envelop, garland

5. **What is the meaning of SEDULOUS?**

   (a) rebellious
   (b) relating to worldly rather than spiritual matters
   (c) containing grease
   (d) persevering in one's duty
   (e) solitary

6. **Which two words are closest in meaning?**

   eluded, mystic, strange, shared, arcane, fabulous

7. **Which two words are most opposite in meaning?**

   pithy, efficient, banal, laudable, wordy, caring

8. **What is the meaning of INSOUCIANT?**

   (a) unnamed
   (b) apathetic
   (c) lacking in restraint
   (d) carefree and lighthearted
   (e) speaking in a slow, deliberate way

9. **Which word in brackets is most opposite in meaning to the word in capitals?**

   VIRULENT (inert, innocuous, flat, tasteful, septic, concealed)

10. **Which word in brackets is closest in meaning to the word in capitals?**

    VITUPERATION (spirit, blame, sparkle, quickness, silence)

11. **Which two words are closest in meaning?**

    abandon, eschew, convey, represent, emanate, spoil

12. **Which word in brackets is most opposite in meaning to the word in capitals?**

    VENT (dismiss, conceal, prohibit, harm, inhibit)

13. **Which two words are most opposite in meaning?**

    imperious, dominate, ancillary, arrogant, dispirited

14. **What is the meaning of POLEMIC?**

    (a) severe criticism
    (b) long and thin
    (c) divided by two opposing groups
    (d) to lay down the law
    (e) uncaring

Continued on page 90

ODD ONE OUT VISUAL EXERCISES

EXAMPLE

In each of the following, decide from the list of figures, which is the odd one out. These exercises are designed to test your agility of mind and creative thinking in appraising a set of figures and choosing the item in each set that does not fit in.

For example:

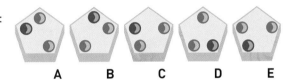

A   B   C   D   E

Answer:
C. All the rest are the same figure rotated. C is a mirror image of the other four figures.

EXERCISE 1

→ Which is the odd one out?

A   B   C

D   E

EXERCISE 2

→ Which is the odd one out?

A   B   C

D   E

EXERCISE 3

→ Which is the odd one out?

A   B

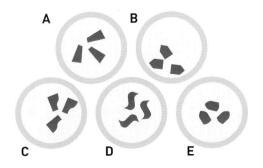

C   D   E

EXERCISE 4

→ Which is the odd one out?

A   B   C

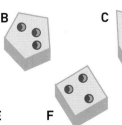
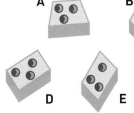

D   E   F   G

Continued on page 91

Turn to pages 144 to 155 for the answers.

VERBAL INTELLIGENCE   *Continued from page 88*

15. Which two words are most opposite in meaning?

idolize, forfeit, confirm, activate, arrest, confess

16. What is the meaning of CHALYBEATE?

[a] milky or grayish quartz
[b] to prepare wine for serving
[c] decorated silver
[d] the jaw or cheek of a pig
[e] containing iron salts

17. Which word in brackets is most opposite in meaning to the word in capitals?

NARROW (long, capricious, small, capacious, adequate)

18. Which word in brackets is most opposite in meaning to the word in capitals?

REPLETE (drain, barren, original, rebuff, attract)

19. Which two words are closest in meaning?

image, analogy, illusion, dotard, chimera, imbroglio

20. Which two words are most opposite in meaning?

singular, propitious, pure, distinct, conventional, phony

21. What is the meaning of DENARY?

(a) the number 10
(b) a denial
(c) starvation
(d) a factory

22. Which is the odd one out?

geyser, fiord, polder, sierra, basinet

23. Which word can be placed in front of these words to make new words?

gear
way
scarf
strong

24. Which word has the same meaning as LEXICON?

parchment, bookcase, textbook, dictionary, chair, desk

25. What is the meaning of OODLES?

scribble, soup, abundance, idiots, nodules

26. What is the name given to a group of LARKS?

(a) parliament
(b) murmuration
(c) exultation
(d) sord
(e) gaggle

27. Which two words are opposite in meaning?

wan, discharge, eject, florid, eddy, rigid

28. Insert a word that means the same as the words outside the brackets.

young mare (_____) lively girl

29. What is a gnomon?

(a) sentry
(b) garden gnome
(c) sundial
(d) hut
(e) cheek

30. Fill in the blanks to find an eight-letter word.

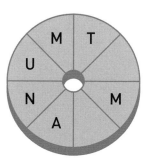

Continued from page 89

ODD ONE OUT VISUAL EXERCISES

EXERCISE 5

→ Which is the odd one out?

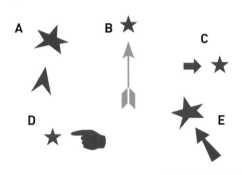

EXERCISE 6

→ Which is the odd one out?

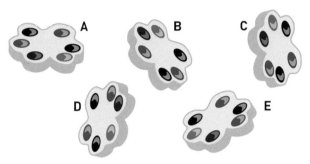

EXERCISE 7

→ Which is the odd one out?

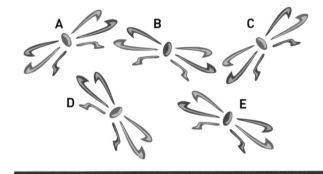

EXERCISE 8

→ Which is the odd one out?

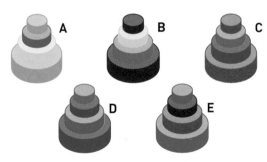

EXERCISE 9

→ Which is the odd one out?

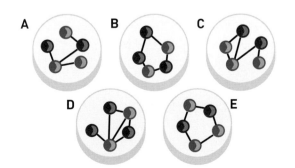

EXERCISE 10

→ Which is the odd one out?

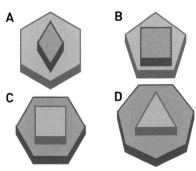

Turn to pages 144 to 155 for the answers.

Exercises to improve the left brain

LETTER/NUMBER LOGIC

→ What number is missing from the grid, and where should it be placed?

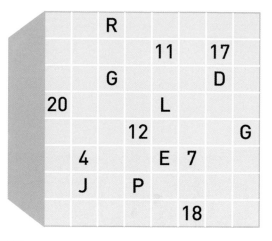

ANALYTICAL PUZZLE

EXAMPLE

It is possible to solve the puzzle below by an analytical and logical method.

How many possible routes are there to get from A to B? It is assumed that to get from A to B by each different route, you will always travel in either an eastern or southern direction.

For example, one way to get from A to B is the following route, shown in red:

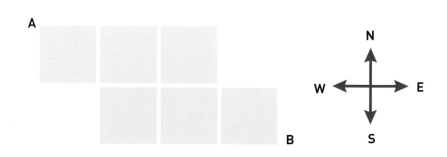

EXERCISE 1

→ How many possible routes are there to get from A to B?

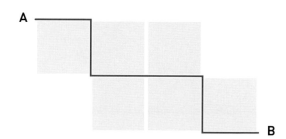

Turn to pages 144 to 155 for the answers.   Continued on page 94

CREATIVE SEQUENCE EXERCISES

→ The following exercises will give the left-brain person practice in identifying a pattern, will enhance their spatial awareness, and will give them the chance to develop their creative talents by drawing the figure that should appear next in the sequence.

For example:

Answer:
The sequence runs brown circle/blue square, blue circle/brown square, and is repeated.

EXERCISE 1

→ Draw the next figure in the sequence.

EXERCISE 2

→ Draw the next figure in the sequence.

EXERCISE 3

→ Draw the next figure in the sequence.

 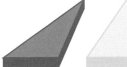 

EXERCISE 4

→ Draw the next figure in the sequence.

Continued on page 95

Turn to pages 144 to 155 for the answers.

ANALYTICAL PUZZLE

Continued from page 92

EXERCISE 2

→ A ball 13 inches in diameter has a 5-inch hole drilled through it. How deep is the hole?

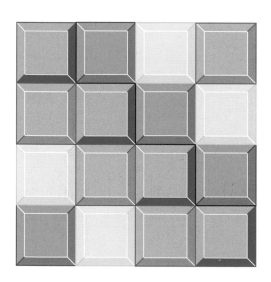

EXERCISE 3

→ How many triangles?

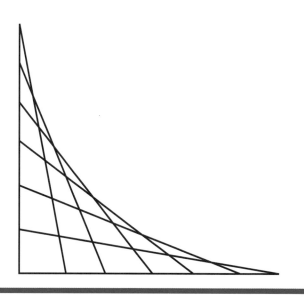

EXERCISE 4

→ How many squares?

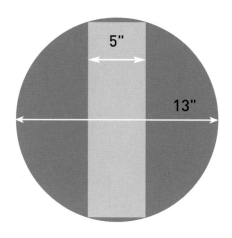

EXERCISE 5

→ How many circles?

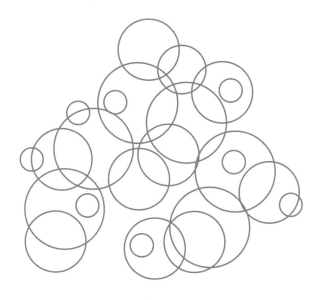

Exercises to improve the right brain

FOR THOSE WITH A LEFT-BRAIN BIAS

Continued from page 93

CREATIVE SEQUENCE EXERCISES

EXERCISE 5

→ Draw the next figure in the sequence.

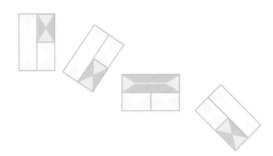

EXERCISE 6

→ Draw the next figure in the sequence.

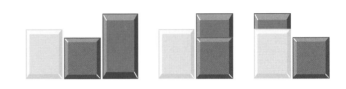

EXERCISE 7

→ Draw the next figure in the sequence.

EXERCISE 8

→ Draw the next figure in the sequence.

EXERCISE 9

→ Draw the next figure in the sequence.

EXERCISE 10

→ Draw the next figure in the sequence.

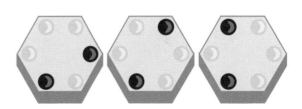

Turn to pages 144 to 155 for the answers.

Exercises to improve the left brain
FOR THOSE WITH A RIGHT-BRAIN BIAS

EXAMPLE

The following is an exercise of 10 matrix puzzles designed to develop your powers of analysis and logic. Take a look across each line, and down each column, and decide from the choice of options, which is the missing square.

For example: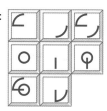

A **B** **C**

D **E** **F**

Answer:
E. Looking across and down, the third square is the first two squares combined.

E

EXERCISE 1

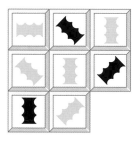

A **B**

C **D**

E **F**

EXERCISE 2

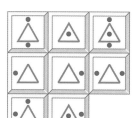

A **B**

C **D**

E **F**

EXERCISE 3

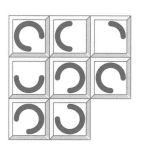

A **B**

C **D**

E **F**

EXERCISE 4

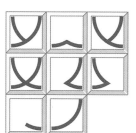

A **B**

C **D**

E **F**

Turn to pages 144 to 155 for the answers.

Continued on page 98

EXAMPLE

Good spatial awareness is essential for everyday life. It not only stimulates mental thought, but it can also have a positive effect on physical performance.

For example:
Which is the odd one out?

A   B   C   D   E

Answer:
B. A is the same as D with blue and red reversed. Similarly, C is the same as E.

EXERCISE 1

  is to as is to

A   B   C   D

EXERCISE 2

 is to as

is to

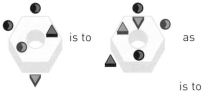

A   B   C   D

EXERCISE 3

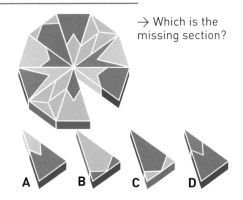

→ Which is the missing section?

A   B   C   D

EXERCISE 4

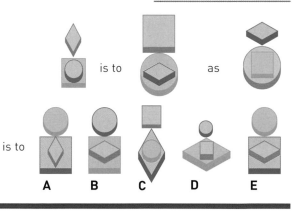

is to   as

is to

A   B   C   D   E

Continued on page 99
Turn to pages 144 to 155 for the answers.

Exercises to improve the left brain

FOR THOSE WITH A RIGHT-BRAIN BIAS

MATRIX PUZZLES

Continued from page 96

EXERCISE 5

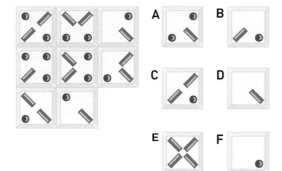

EXERCISE 6

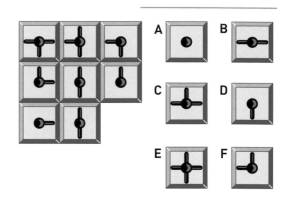

EXERCISE 7

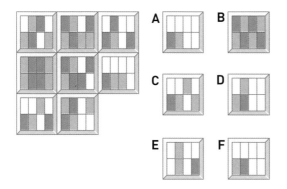

EXERCISE 8

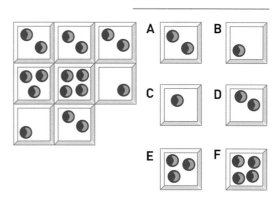

EXERCISE 9

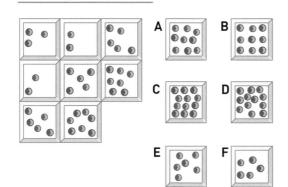

EXERCISE 10

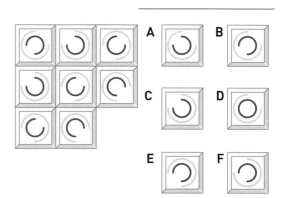

Continued from page 97

SPATIAL AWARENESS

EXERCISE 5

→ Which ball is missing from the circle?

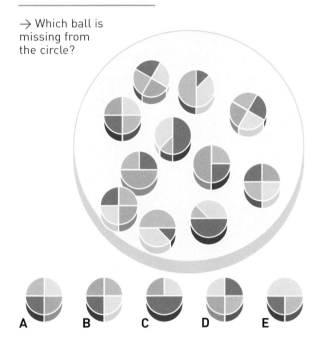

EXERCISE 6

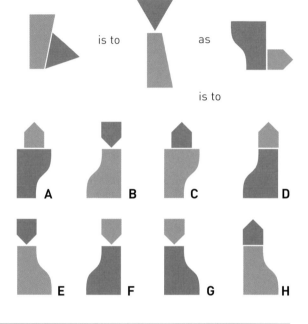

is to — as

is to

EXERCISE 7

→ Which is the odd one out?

EXERCISE 8

→ Which is the odd one out?

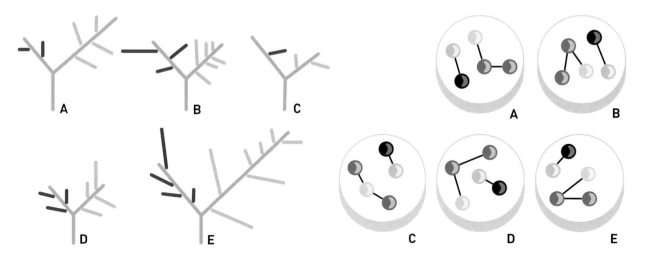

Continued on page 101

Turn to pages 144 to 155 for the answers.

Exercises to improve the left brain
FOR THOSE WITH A RIGHT-BRAIN BIAS

MAGIC SQUARES

EXAMPLE

The magic-square puzzles that follow are designed to exercise your ability to concentrate on one specific task. Some of them involve a great deal of number crunching and juggling with figures. Patience, determination, and analytical thought are required to solve them. Magic-square puzzles were developed by the ancient Chinese and consist of an array of numbers in which all rows, columns, and diagonals add up to the same total. The first ever magic square is known as the Lo-shu and, according to Chinese legend, is said to have appeared to the mythical Emperor Yu on the back of a tortoise. The illustration shown here consists of arrangements of beads, blue for even numbers and yellow for odd, in which the number of beads on each horizontal, vertical, and diagonal line add up to 15. Here's how it works:

The Loh River (Lo-shu) 15

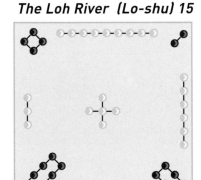

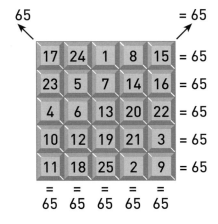

Magic 65
The puzzle uses numbers 1 to 25 only one time each and all lines total 65.

EXERCISE 1

→ This is an anti-magic square where no lines, horizontal, vertical, or corner to corner, add up to 34. By changing the position of just four of the numbers, can you create a magic square so that each horizontal, vertical, and corner-to-corner line totals 34?

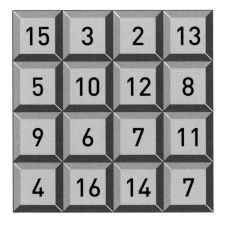

Turn to pages 144 to 155 for the answers.   Continued on page 102

Continued from page 99

SPATIAL AWARENESS

EXERCISE 9

→ Which is the odd one out?

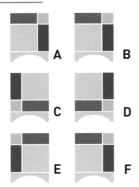

EXERCISE 10

→ Which circle should replace the question mark?

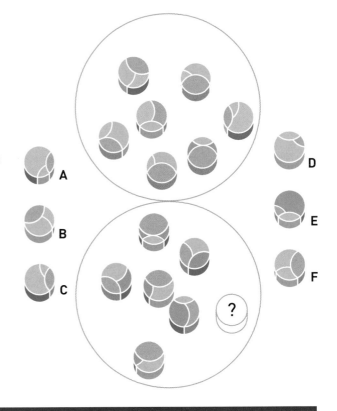

EXERCISE 11

→ Which is the odd one out?

EXERCISE 12

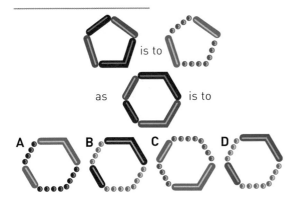

is to

as   is to

EXERCISE 13

→ Which is the odd one out?

Continued on page 103

Turn to pages 144 to 155 for the answers.   101

MAGIC SQUARES

Continued from page 100

EXERCISE 2

→ Insert the remaining numbers from 1 to 19 to complete the pattern in such a way that all 15 connected straight lines —horizontals and diagonals—add up to 51. For example: A + B + C = 19; A + E + J + O + S = 19. Pink circles should contain even numbers, and blue circles should contain odd numbers.

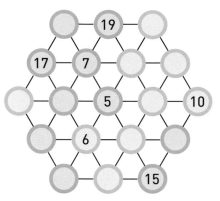

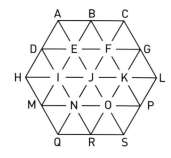

EXERCISE 3

→ Change four numbers in the left-hand grid with four numbers in the right-hand grid, so that in both grids each horizontal, vertical, and corner-to-corner line adds up to 65.

14	3	11	13	24
19	23	7	10	6
20	12	1	17	15
4	22	25	9	5
8	2	21	18	16

25	10	3	6	21
22	15	19	8	4
11	9	13	17	12
2	16	7	14	24
5	18	23	20	1

EXERCISE 4

→ Insert the remaining numbers so that each grid contains the numbers 1 to 25 only one time each, and every horizontal, vertical, and corner-to-corner line totals 65.

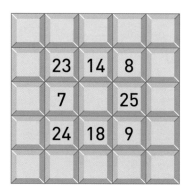

Continued on page 104

Continued from page 101

SPATIAL AWARENESS

EXERCISE 14

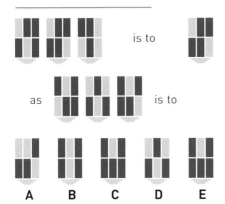

is to

as

is to

A   B   C   D   E

EXERCISE 15

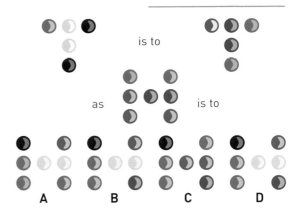

is to

as

is to

A   B   C   D

EXERCISE 16

→ Which is the odd one out?

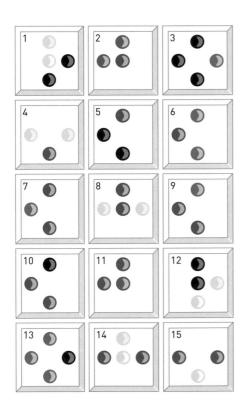

EXERCISE 17

→ Which is the missing tile?

A   B

C   D

E   F

EXERCISE 18

is to

as

is to

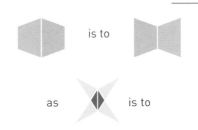

A   B   C   D   E

Continued on page 105

MAGIC SQUARES

Continued from page 102

EXERCISE 5

→ Put the pieces together to form a magic square where each horizontal, vertical, and corner-to-corner line totals 111.

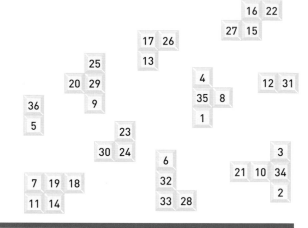

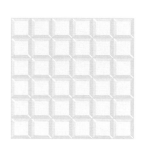

EXERCISE 6

→ Insert the remaining numbers from 1 to 36 so that each row, column, and corner-to-corner line adds up to 111.

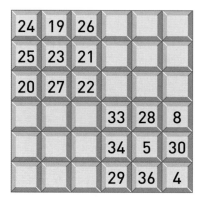

EXERCISE 7

→ Place the numbers 1 to 25 to form a magic square where each horizontal, vertical, and diagonal line totals 65. The prime numbers 2, 3, 5, 7, 11, 13, 17, 19 and 23 should be placed in the red shaded squares only.

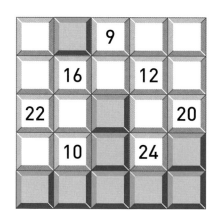

Continued from page 103

SPATIAL AWARENESS

EXERCISE 19

→ Which is the odd one out?

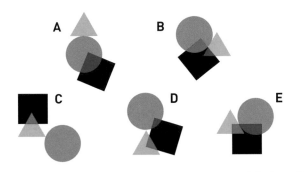

EXERCISE 20

→ Which is the odd one out?

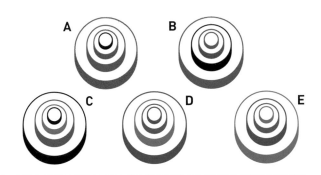

EXERCISE 21

→ Which is the odd one out?

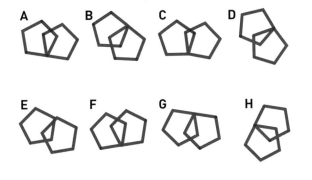

EXERCISE 22

→ This sequence requires a higher degree of lateral thinking as well as sequential analysis and is therefore included as an exercise to improve the right brain.

What comes next?

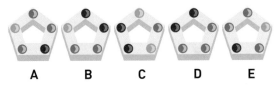

EXERCISE 23

→ Which is the odd one out?

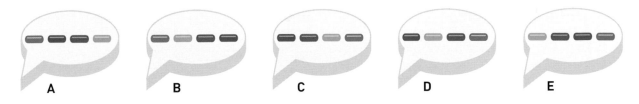

Turn to pages 144 to 155 for the answers.

Exercises to improve the left brain

FOR THOSE WITH A RIGHT-BRAIN BIAS

WORD CHANGES

EXAMPLE

The following exercises are designed to test your powers in understanding word meanings, and your ability to think quickly under pressure.

For example:

Not all vast publishing is commercial and conglomerate.

Change around "vast" and "commercial" to make:

Not all commercial publishing is vast and conglomerate.

→ In each of the following, choose the **two** words that if swapped around would make the sentence comprehensible:

1. Education is the grade whereby one acquires a higher method of prejudices.

2. Tsunamis start out as barely noticeable deepwater eruptions caused by underwater earthquakes or volcanic ripples.

3. An effective business telephone style can only be achieved by personally reviewing the practices we constantly develop as time goes by.

4. Prior to any job, the interviewer must have a clear picture of what he or she is looking for in the candidate being interviewed in relation to the requirement of the interview.

→ In the following, choose the **three** words that if swapped around would make the sentence comprehensible:

5. Today, buffs are drinking the wines they enjoy, irrespective of whether the so-called wine values still embrace old-fashioned people.

→ In each of the following, choose the **four** words that if swapped round would make the sentence comprehensible:

6. Early afternoon is usually set aside to accomplish set work; late morning appointments make it more difficult to complete the day's routines.

7. Libraries have comprehensive reference facilities at their disposal, and even if the material is not available from their index information, they will know who you should contact.

→ In the following, swap around only **three** words:

8. From the lake high above the fairway, the distance across to the tee looks shorter than it is, as distances invariably do over water.

→ Now swap **four** words again:

9. Truly surreal writing can transcend its own heights to rise to awful inadequacies.

→ Now swap around **six** words:

10. The generation of popular novelists in the late nineteenth century spawned a spread of tenth-rate elementary education.

→ This test was devised over 50 years ago to encourage creative thought. There are several solutions to the problem; however, just one stands out as being eminently superior to the others.

You have at your disposal a candle, a book of matches, and a box of thumbtacks. With these items and nothing more, you must attach the candle to a wooden door in such a way that it throws out sufficient light to read by.

What is the most efficient way to achieve this task?

Turn to pages 144 to 155 for the answers.

Exercises to improve the left brain

FOR THOSE WITH A RIGHT-BRAIN BIAS

MATHEMATICAL CALCULATION: BRAIN STRAIN

EXAMPLE

In each of the following, complete the grid so that all the calculations are correct reading across and down.

All numbers to be inserted are below 10.

Here is a sample grid, filled in, to show you how it works.

5	x	3	–	9	=	6
+		x		+		–
4	–	2	+	1	=	3
÷		–		–		+
3	x	4	÷	6	=	2
=		=		=		=
3	–	2	+	4	=	5

EXERCISE 1

EXERCISE 2

Continued on page 110

POOL AND SNOOKER

EXAMPLE

A top pool or snooker player requires natural talent, persistence, and focus, but flair, creativity, and spatial awareness are absolutely essential.

In the following puzzles you have one ball, the red ball, on the table and you need to pocket it with the black ball. Several of your opponent's balls, the blue balls, are on the table, and you must figure out how to shoot your last remaining red ball without touching any of the blue balls in the process. In each case, work out how you can strike the black ball to travel around the table and knock the red ball into a pocket. In each instance, what is the minimum number of sides, or cushions, of the table the black ball needs to rebound off to achieve this task?

For example:

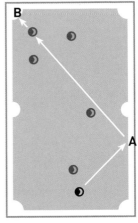

→ By rebounding the black ball off the cushion at point A, it strikes the red ball and places it into pocket B.

EXERCISE 1

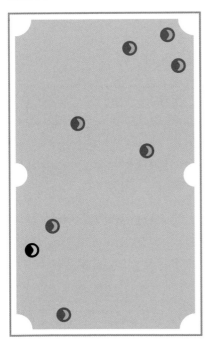

EXERCISE 2

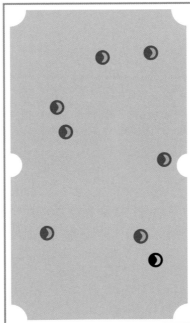

EXERCISE 3

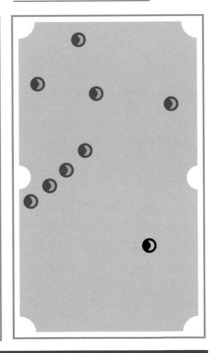

Turn to pages 144 to 155 for the answers.

MATHEMATICAL CALCULATION: BRAIN STRAIN

Continued from page 108

EXERCISE 3

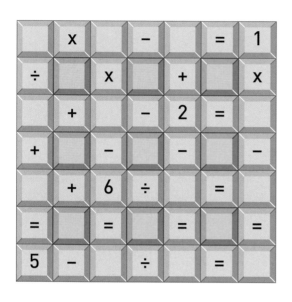

EXERCISE 4

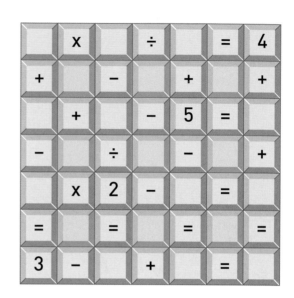

EXERCISE 5

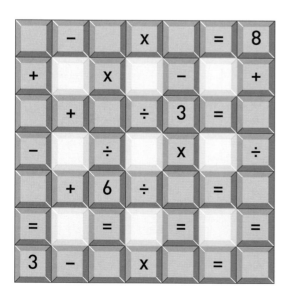

EXERCISE 6

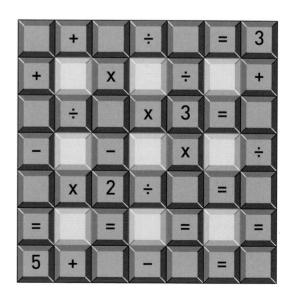

EXAMPLE

These puzzles require some logic—a left-brain function—but they also involve a high level of visual awareness and creative thinking—right-brain functions.

Here's an example of how this logic puzzle works:
What comes next in the sequence?

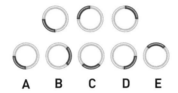

A B C D E

Answer: D. The red arc moves 90 degrees clockwise at each stage.

EXERCISE 1

→ What comes next?

A B C D E

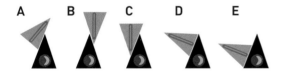

EXERCISE 2

→ What comes next?

A B C D E

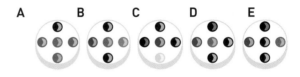

EXERCISE 3

→ Which is the missing tile?

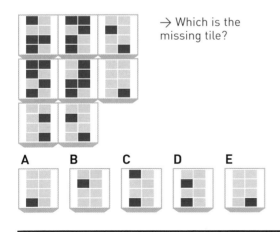

A B C D E

EXERCISE 4

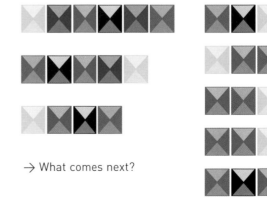

→ What comes next?

A

B

C

D

E

Continued on page 113

Turn to pages 144 to 155 for the answers.

Exercises to improve the left brain
FOR THOSE WITH A RIGHT-BRAIN BIAS

MONOGRAMS

EXAMPLE

Invented by the authors, monograms are visual representations of a word or words formed by the circular arrangement of the letters of the alphabet in the grid shown. The

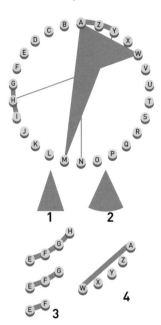

best examples form patterns that can, in some way, be related to the word or words represented. The patterns are formed in the following way. For example, the word HIGHWAYMAN is plotted out starting at the first letter, H, and lines are drawn from letter to letter, such as H to I, I to G, G to H, H to W, and W to A. When all the lines have been drawn, all triangular areas of the pattern produced are filled in if bounded by straight lines (1), but are not filled in if bounded by one or more curved lines (2). In the event of adjoining letters of up to three letters (3), a thick line is drawn on the circumference. If, however, the move between letters exceeds three moves (4), a line is drawn to link the first and last letters. In the event of double letters, just one of the double letters is visited.

Stage 1: Draw connecting lines.

Stage 2: Fill in all areas completely bounded by straight lines.

EXERCISE 1

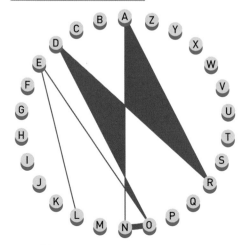

Clue: Figure of the Renaissance?

EXERCISE 2

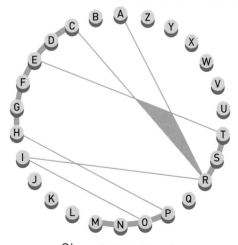

Clue: Built by Napoleon

Continued on page 114

Continued from page 111

CREATIVE LOGIC

EXERCISE 5

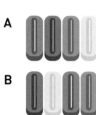

→ What comes next?

A
B
C
D
E

EXERCISE 6

→ What square should replace the question marks?

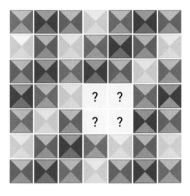

A B C D E

EXERCISE 7

→ What comes next?

A   B   C

D   E   F

EXERCISE 8

→ What comes next?

A   B   C

D   E

Continued on page 115

Turn to pages 144 to 155 for the answers.

Exercises to improve the left brain
FOR THOSE WITH A RIGHT-BRAIN BIAS

MONOGRAMS

Continued from page 112

EXERCISE 3

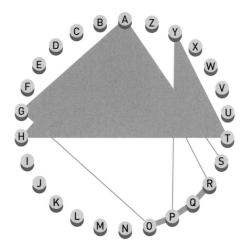

Clue: Greek philosopher

EXERCISE 4

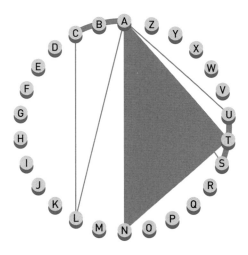

Clue: Fourth-century Christmas saint

EXERCISE 5

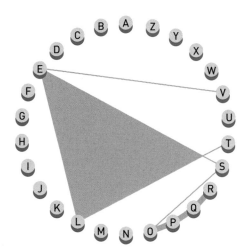

Clue: President born in New York

EXERCISE 6

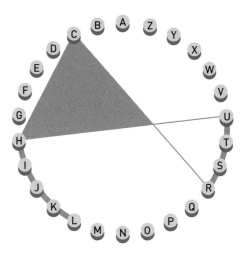

Clue: 20th-century English statesman

Turn to pages 144 to 155 for the answers.

Continued on page 116

Continued from page 113

CREATIVE LOGIC

EXERCISE 9

→ What comes next?

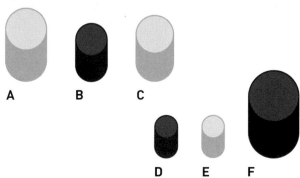

EXERCISE 10

→ What comes next?

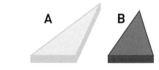

EXERCISE 11

→ What comes next?

A   B   C

D   E   F

EXERCISE 12

→ What comes next?

A   B   C

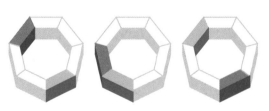

D   E

Continued on page 117

Turn to pages 144 to 155 for the answers.

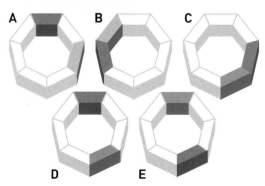

MONOGRAMS

Continued from page 114

EXERCISE 7

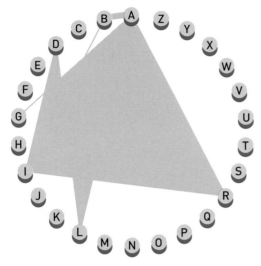

Clue: Italian soldier who was fond of biscuits

EXERCISE 8

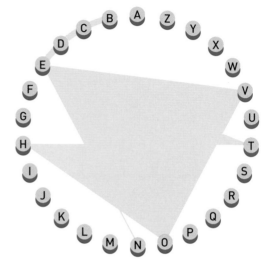

Clue: German composer

EXERCISE 9

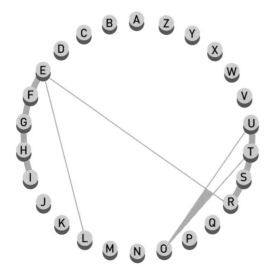

Clue: It's 1,050 feet high in Paris

EXERCISE 10

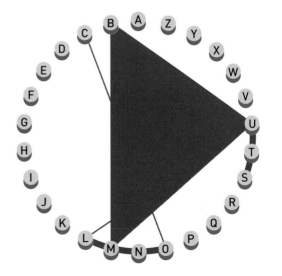

Clue: Explorer born in Genoa

Continued from page 115

CREATIVE LOGIC

EXERCISE 13

→ Which is the missing tile?

A   **B**   **C**   **D**   **E**

EXERCISE 14

→ What comes next?

A   **B**

C   **D**   **E**

EXERCISE 15

→ What comes next?

A   **B**   **C**

D   **E**

EXERCISE 16

→ What comes next?

Exercises to improve the left brain

FOR THOSE WITH A RIGHT-BRAIN BIAS

WORDS

EXAMPLE

In exercise 1, what word should be placed inside the brackets that means the same as the two words outside the brackets? For example:

To move heavily (lumber) Timber in store

In exercise 2, what is the name given to the group specified. For example:

Is a group of larks [a] an exultation [b] a raft [c] a rookery or [d] an ostentation?   Answer: A

EXERCISE 1

1. Two-wheeled carriage (_____) Musical performance

2. Fasten by cable (_____) Tract of land

3. Barren tract of land (_____) Forsake

4. To droop (_____) Iris

5. Cleaning woman (_____) Reduce to carbon

6. Whale fat (_____) Weep copiously

7. To hoax (_____) Young goat

8. Loud sound (_____) A young sea trout

9. Arrest (_____) Squeeze tightly

10. Confer knighthood (_____) Create new soundtrack

EXERCISE 2

→ What is the name given to a group of:

1. **Whales**
    (a) grist
    (b) mute
    (c) cede
    (d) pod

2. **Ravens**
    (a) simplicity
    (b) unkindness
    (c) building
    (d) tiding

3. **Antelopes**
    (a) herd
    (b) dray
    (c) pack
    (d) gang

4. **Hermits**
    (a) caste
    (b) sloth
    (c) skulk
    (d) observation

5. **Toads**
    (a) knot
    (b) bale
    (c) host
    (d) nest

6. **Starlings**
    (a) collection
    (b) swarm
    (c) murmuration
    (d) chattering

7. **Princes**
    (a) school
    (b) company
    (c) state
    (d) clique

8. **Ducks**
    (a) paddling
    (b) plump
    (c) skein
    (d) company

EXERCISE 1

→ Find the hexagon that fills the space correctly.

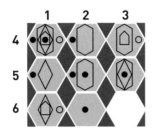

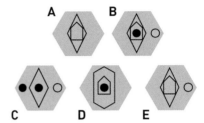

EXERCISE 2

→ Find the hexagon that fills the space correctly.

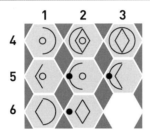

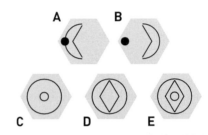

EXERCISE 3

→ Find the hexagon that fills the space correctly.

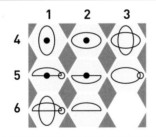

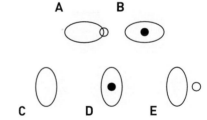

EXERCISE 4

→ Find the hexagon that fills the space correctly.

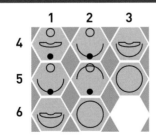

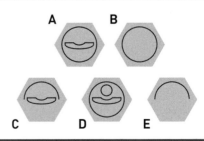

EXERCISE 5

→ Find the hexagon that fills the space correctly.

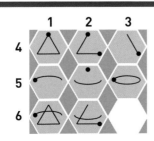

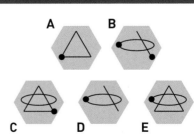

Continued on page 121   Turn to pages 144 to 155 for the answers.

Exercises to improve the left brain

FORE WORDS

EXAMPLE

Place the same word in front of each of these groups of words to make new words.

For example: (_____) blind   Answer: sun
                      light
                      shade
                      burn
                      deck

1.   (_____) stripe
             cushion
             table
             hole
             head

2.   (_____) ball
             bridge
             hold
             lights
             print

3.   (_____) copy
             nosed
             headed
             lined
             hit

4.   (_____) boat
             like
             kin
             burglar
             call

5.   (_____) field
             row
             snake
             borer
             flower

6.   (_____) bug
             digger
             dust
             fish
             brick

7.   (_____) hole
             handle
             hour
             fully
             kind

8.   (_____) legs
             boat
             ways
             bow
             haul

9.   (_____) ton
             park
             sick
             port
             wash

10. (_____) back
             bridge
             string
             bar
             able

11. (_____) ship
             fare
             monger
             den
             lock

12. (_____) head
             like
             child
             parent
             daughter

13. (_____) did
             ton
             dor
             not
             on

14. (_____) dust
             board
             shell
             light
             lit

15. (_____) ness
             speaker
             mouth
             hailer
             voiced

16. (_____) shine
             day
             lounge
             shade
             dry

17. (_____) land
             stay
             stream
             frame
             sail

18. (_____) ping
             room
             dance
             shoes
             per

19. (_____) keeper
             mark
             rest
             end
             store

20. (_____) fast
             back
             forth
             down
             up

Continued from page 119

HEXAGONS

EXERCISE 6

→ Find the hexagon that fills the space correctly.

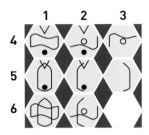

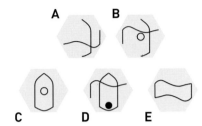

EXERCISE 7

→ Find the hexagon that fills the space correctly.

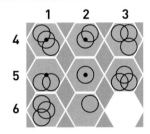

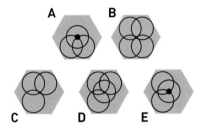

EXERCISE 8

→ Find the hexagon that fills the space correctly.

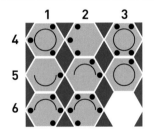

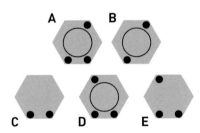

EXERCISE 9

→ Find the hexagon that fills the space correctly.

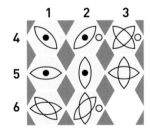

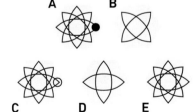

EXERCISE 10

→ Find the hexagon that fills the space correctly.

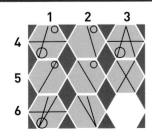

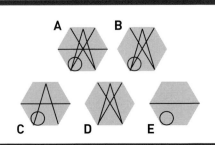

Turn to pages 144 to 155 for the answers.

Exercises to improve the left brain

FOR THOSE WITH A RIGHT-BRAIN BIAS

TECHNICAL

EXERCISE 1

→ What are these letters of the Greek alphabet?

Choose from:

Sigma	Rho
Beta	Delta
Pi	Iota
Eta	Theta
Upsilon	Gamma

$$H \quad \Gamma \quad \Delta \quad \Theta \quad Y$$

EXERCISE 2

→ What is the height of a cliff if a stone is dropped and it takes 3 seconds to hit the sea?

EXERCISE 3

→ What is the formula for the volume of a sphere?

EXERCISE 4

→ Where is the center of gravity of a flat lamina sheet? Will it be inside the sheet or outside?

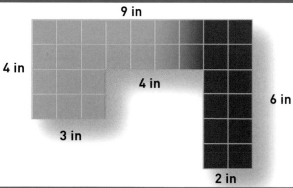

9 in

4 in

4 in

3 in

6 in

2 in

Turn to pages 144 to 155 for the answers.

Continued on page 124

EXERCISE 1

→ Which hexagon replaces the ?

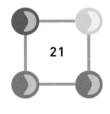

EXERCISE 2

→ Which colors should replace the ?

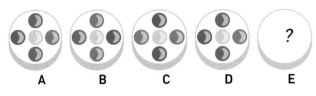

EXERCISE 3

→ Which circle is nearest in content to A?

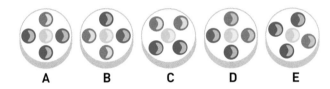

EXERCISE 4

→ Yellow, green, pink, and blue have a value of 10-5-3-2, but not necessarily in that order. What is the true order?

EXERCISE 5

→ Which color should replace the ?

Continued on page 125

Turn to pages 144 to 155 for the answers.

EXERCISE 5

→ The Beaufort wind scale goes from force 0 to 12.

Which force is 41 to 47 knots, has high waves, streaks of foam following the direction of the wind, strong gale winds, and an average wave height of 25 feet?

EXERCISE 6

→ A large plate has to be completely covered by five smaller plates. What is the minimum diameter of the smaller plates expressed as a percentage of the larger?

Percent

(a) 58.8
(b) 59.8
(c) 60.8
(d) 61.8
(e) 62.8

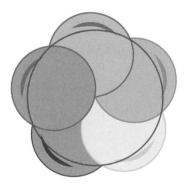

EXERCISE 7

→ Which compass direction is shown?

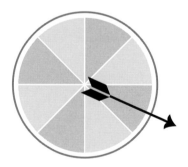

EXERCISE 8

→ What are the meanings of these mathematical symbols?

Choose from:

1. universal set
2. is equal to
3. plus or minus
4. is much less than
5. is greater than
6. there exists
7. infinity
8. is less than
9. square root of
10. imaginary number
11. is not equal to
12. mean value of

$$\pm \quad > \quad \neq \quad \sqrt{} \quad \propto$$

Continued on page 126

Continued from page 123

COLORS, CIRCLES, SQUARES, PENTAGONS

EXERCISE 6

→ Which is the odd one out?

EXERCISE 7

→ Which square should replace the ?

EXERCISE 8

→ Analogy

is to

Then

is to

EXERCISE 9

→ Which is the odd one out?

EXERCISE 10

→ Which pentagon should replace the ?

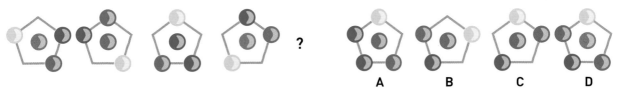

Turn to pages 144 to 155 for the answers.

Continued from page 124

EXERCISE 9

→ How do you change Celsius to Fahrenheit and Fahrenheit to Celsius, and which two temperatures are the same in both systems?

EXERCISE 10

→ If dry table salt falls freely, it forms a cone shape. What is the angle of repose?

Angle

(a) 25 degrees
(b) 33 degrees
(c) 35 degrees
(d) 40 degrees
(e) 43 degrees

EXERCISE 11

→ Which is the strongest beam?

(a) 2 x 4 in
(b) 1½ x 5 in
(c) 3 x 3 in
(d) 1 x 6 in
(e) 1¾ x 4½ in

A   B   C   D   E

EXERCISE 12

→ If a ball is thrown into the air it traces out a curve. What is the name of this curve?

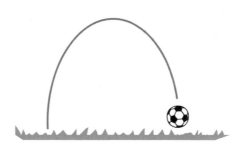

(a) catenary
(b) cardioid
(c) conic
(d) parabola
(e) hyperbola
(f) equilateral

Continued on page 128

EXERCISE 1

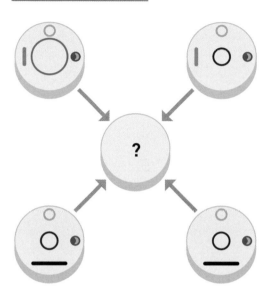

→ Each line and symbol that appears in the four outer circles above is transferred to the center circle according to these rules:

**If a line or symbol occurs in the outer circles
once: it is transferred
twice: it is possibly transferred
three times: it is transferred
four times: it is not transferred**

→ Which of the circles shown below should appear at the center of the diagram above?

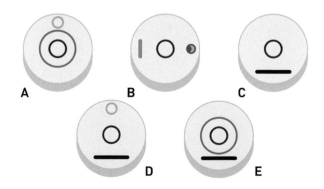

EXERCISE 2

→ Each of the nine squares in the grid marked 1A to 3C should incorporate all the lines and symbols that are shown in the squares of the same letter and number immediately above and to the left. For example, 2A should incorporate all the lines and symbols that are in 2 and A.

One of the squares is incorrect. Which one is it?

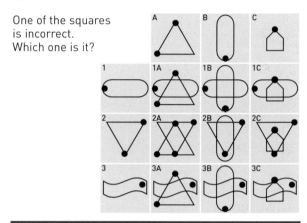

EXERCISE 3

→ Which is the odd one out?

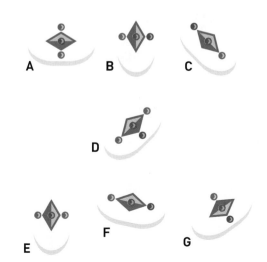

Continued on page 129

Turn to pages 144 to 155 for the answers.

TECHNICAL

Continued from page 126

EXERCISE 13

→ What is the volume of the dunce's hat?

(a) 256
(b) 286
(c) 316
(d) 346
(e) 376

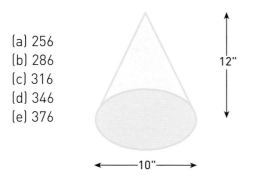

EXERCISE 14

→ If two dice are thrown, what is the possibility of scoring 7 total?

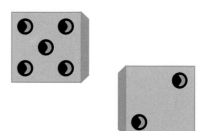

EXERCISE 15

→ What is the height of line AB?

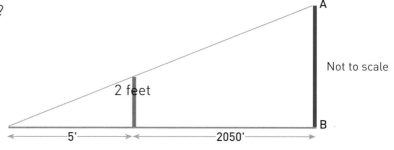

EXERCISE 16

→ Five men are building a brick wall together. At their respective speeds, if they were working alone:

Man one would take 3 hours
Man two would take 6 hours
Man three would take 2 hours
Man four would take 10 hours
Man five would take 4 hours

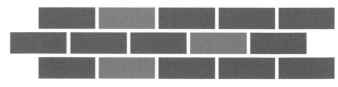

→ How long did they take working together?

Continued on page 130

Continued from page 127 ## SYMBOLS, GRIDS, AND OTHERS

EXERCISE 4

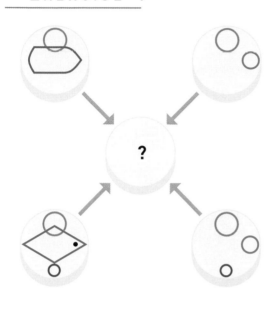

→ Each line and symbol that appears in the four outer circles above is transferred to the center circle according to these rules:

**If a line or symbol occurs in the outer circles
once: it is transferred
twice: it is possibly transferred
three times: it is transferred
four times: it is not transferred**

→ Which of the circles shown below should appear at the center of the diagram above?

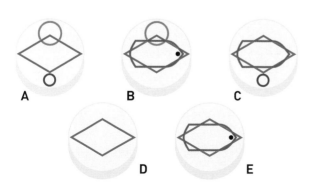

EXERCISE 5

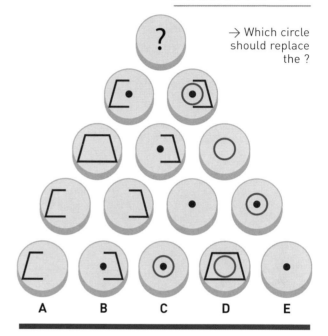

→ Which circle should replace the ?

EXERCISE 6

Each of the nine squares in the grid marked 1A to 3C should incorporate all the lines and symbols that are shown in the squares of the same letter and number immediately above and to the left. For example, 2A should incorporate all the lines and symbols that are in 2 and A.

One of the squares is incorrect. Which one is it?

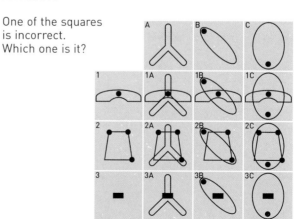

Exercises to improve the left brain

FOR THOSE WITH A RIGHT-BRAIN BIAS

TECHNICAL

Continued from page 128

EXERCISE 17

→ What is the decimal value of this binary number?

11111.111

EXERCISE 18

→ What is the value of an angle in a pentagon?

EXERCISE 19

→ A piece of machinery weighs 20 pounds plus three-eighths of its total weight. What does it weigh?

EXERCISE 20

→ Which planet has 18 moons?

Exercises to improve the right brain
FOR THOSE WITH A LEFT-BRAIN BIAS

CREATIVE NUMBER PUZZLES

→ No mathematical knowledge is required to solve these two somewhat zany number puzzles, just a high degree of creative thought in seeking out the unlikely and unexpected.

EXERCISE 1

1, 11, 21, 1112, 3112, 211213 , 312213 , 212223 , 114213 ,

31121314, 41122314, 31221324, 21322314, ?

→ What number should replace the question mark?

EXERCISE 2

9, 18, 23, 26, 29, 46, ?, 52

→ What number should replace the question mark?

ALTERNATING PUZZLE

→ Below are five glasses of orange soda and five empty glasses.

1   2   3   4   5   6   7   8   9   10

→ How many glasses do you need to pick up so there are still 10 glasses and full and empty glasses alternate?

LOGIC

1. Expand $(2x - y)^2$

   Select from

   (a) $-4x^2 - 4xy + y^2$
   (b) $-4x^2 + 4xy + y^2$
   (c) $4x^2 - 4xy - y^2$
   (d) $4x^2 - 4xy + y^2$
   (e) $4x^2 + 4xy + y^2$

2. What value weight must be placed at the ? to make the scales balance

3. What numbers should replace the ??

   9 16 ? 64 729

   8 27 ? 243 128

4. Which three digit number comes next?

   394 – 041 – 424 – 344 - ?

5. Simplify

   $\dfrac{7}{41} \times \dfrac{123}{28} = x$

6. Find the value of ?

   4½, 33, 8, 25½, 11½, 18, 15, ?

7. If

   7 x 6 = 46
   4 x 5 = ?

8. Simplify

   $6 - 4 \times 10 + 8 \div 2 - 12 = x$

9. Simplify

   $\dfrac{6}{1\frac{2}{3} - \frac{1}{2}} = x$

10. Which of these is a BABOON?

    (a) MANDREL
    (b) MANDRAKE
    (c) MANDIRA
    (d) MANDRILL
    (e) MANDRIL

11. Insert a word which means the same as the words outside the brackets

    FISH (_____) COLOR

12. What is a GREMLIN?

    (a) fish
    (b) bird
    (c) goblin
    (d) clown
    (e) flower
    (f) insect

13. Which is the odd one out?

    (a) hurricane
    (b) monsoon
    (c) doldrums
    (d) mistral
    (e) typhoon

14. Which one has letters which you cannot arrange into a word?

    (a) nhwci
    (b) rgwei
    (c) iywnd
    (d) riwep
    (e) wnisg

15. Which word means the same as SAGACIOUS?

    (a) violent
    (b) effete
    (c) discerning
    (d) lithe
    (e) ugly

16. If TRADE TERM is to AMORTIZE, then MILITARY TERM is to

    (a) nubile
    (b) nautch
    (c) grenadine
    (d) defilade
    (e) mezzotint

17. Which word can be placed in front of these words to make new words?

    (_____)   gear
                stone
                way
                scarf
                long

18. Arrange the following letters to form two five-letter words

    AETORRSZ

    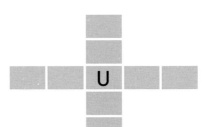

Exercises to improve the right brain

FOR THOSE WITH A LEFT-BRAIN BIAS

CREATIVITY: THE TRICK DONKEYS

→ Sam Loyd was a creative nineteenth-century genius who compiled thousands of original puzzles, many of which are still famous today.

Born in Philadelphia in 1841, Loyd was only 17 when he invented one of his most famous puzzles, *The Trick Donkeys*. The puzzle was purchased by circus owner P.T. Barnum, who sold it for one dollar each at his circus performances as *P.T. Barnum's Trick Donkeys*.

The object of the puzzle is to cut out the three pieces along the dotted lines and then arrange the pieces so that the two riders are riding the two donkeys.

It's a deceptive puzzle—even if you are shown the answer, you may have forgotten how the puzzle is solved the next time.

Try to visualize in your mind how the puzzle can be solved without cutting out the three pieces. You will find that it is a wonderful exercise in creative thinking.

Turn to pages 144 to 155 for the answers.

MATHEMATICAL LOGIC

1. What number should replace the question mark?

6, 23, 7½, 21, 9, 19, 10½, 17, ?

2. Find the value of x.

$-17 - (-17) - (-17) = x$

3. Find the value of x.

$6 - 17 \times 2 - 14 \div 2 = x$

4. Find the value of x.

$\frac{64}{10} \div \frac{8}{5} = x$

5. What is the value of angle A?

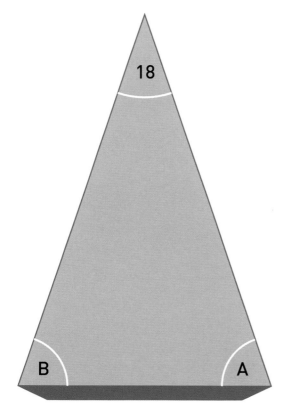

6. Replace the question mark with a number.

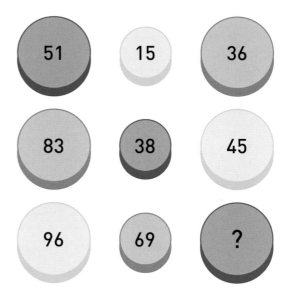

7. Find x.

$54^2 - 53^2 = x$

8. What number should replace the question mark?

17, $8^{7}/_{8}$, $^{3}/_{4}$, $-7^{3}/_{8}$, ?

9. What number should replace the question mark?

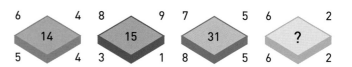

10. Simplify

$(3 + 2) - (16^2 - 9^2)$

Exercises to improve the right brain

FOR THOSE WITH A LEFT-BRAIN BIAS

THOUGHT-PROCESSING EXERCISE

→ The following exercise is based on Gestalt and Jackson's Test of Divergent Ability, which required the subject to name as many new uses as possible for an object such as a comb or a piece of string.

  In this exercise you are required to name as many uses as possible for a brick. Allow yourself 6 minutes to write up to 11 suggestions such as the example below.

1. Use it to prop open your garage door on a windy day.

2. _____

3. _____

4. _____

5. _____

6. _____

7. _____

8. _____

9. _____

10. _____

11. _____

12. _____

Scoring:
Award yourself the following scores:
2 points for any good, original, or useful answer.
1 point for not-so-original answers that, nevertheless, are considered to be a good attempt.
0 points for completely impractical answers.
0 points for any antisocial answers, such as breaking a plate glass window or hitting someone over the head.

Analysis:
18 to 24 points: A highly creative and imaginative effort
13 to 17 points: Above average
7 to 12 points: Average
0 to 6 points: You need more practice

Now try repeating the exercise with other household everyday objects, such as a bucket or elastic band.

Exercises to improve the left brain

FOR THOSE WITH A RIGHT-BRAIN BIAS

MATHEMATICAL CALCULATION

1. You have a frying pan that takes two pieces of bread at a time. You wish to fry three slices, each one on both sides. Each slice takes 20 seconds for each side. You can fry them all in 80 seconds by doing two pieces together, and then the third.

   Find a more efficient method.

2. A man went into a clothing store where there was a sale. He was offered a 5 percent discount for using cash, a 10 percent discount for being a regular customer, and a 25 percent discount as the sale price.

   Which order of discounts gives the largest total discount?

3. If to my age there added be, One half, one third, and three times three, Six score and 10, the sum you'll see.

   Pray find out what my age may be?

4. A car travels 30 miles in 60 minutes, and the identical return trip is done in 30 minutes.

   What was the average speed for the two journeys?

5. Three men own a business. A owns 60 percent, B owns 35 percent, C owns 5 percent. They make a profit of $350 in the year.

   How much does each receive?

6. At the zoo there were 117 quadrupeds and 57 birds.

   How many heads and legs were there?

7. A family has three soldiers. One comes home every five days. One comes home every four days. One comes home every three days.

   In how many days will all three meet?

8. The sum of five numbers, omitting each of the numbers in turn is:

   36 – 38 – 41 – 34 – 31

   What are the five numbers?

9. A man caught a fish. It weighed ⅚ lb, plus ⅚ of its own weight.

   How much did it weigh?

10. Some aliens visited Earth. 70 percent had one eye, 75 percent had one arm, 80 percent had one leg, and 85 percent had one ear.

    What percentage at least must have had all four?

11. Twice six are eight of us
    Six are but three of us
    Nine are but four of us
    What can we possibly be?
    Would you like to know more of us?
    I'll tell you more of us
    Twelve are but six of us
    Five are but four of us
    Now do you see?

    What does this text mean?

12. A wine shop has wine at $9.50 a bottle and at $5.50 a bottle.

    How many bottles of wine of each must the owner mix together to sell wine at $7.90 a bottle and still make the same profit?

13. How many ways can different teams of 11 soccer players be chosen from a squad of 17?

14. A train takes 3 seconds to enter a 1-mile tunnel in total.

    If it was traveling at 120 miles per hour, how long would it take to pass completely through the tunnel?

15. If 5 x 4 = 30.

    What will one-quarter of 20 be?

16. Which number, when added separately to 100 and 164, will make them into square numbers?

17. A company gave $3,395 to its employees as a Christmas bonus. There were more than 50 but less than 100 employees. Each received an equal amount.

    How many employees were there, and how much did each receive?

18. At an eating contest, the winner ate an average of 22 hot dogs at the first 10 sittings. After a further 20 sittings, his average increased to 34 hot dogs.

    What was the average for the last 20 sittings?

19. In 1932 I was as old as the last two digits of my birth year. My grandfather said that it applied to him as well.

    What were the ages of both of us, and when were we born?

Exercises to improve the right brain
FOR THOSE WITH A LEFT-BRAIN BIAS

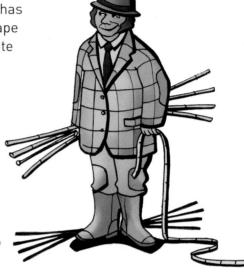

CREATIVE PUZZLES: PUDDLED

→ After a particularly heavy rainstorm, a small puddle has appeared in your backyard. All you have available is a tape measure and some garden stakes. How can you calculate the area of the puddle?

THOUGHT-PROCESSING EXERCISE: SHAPES

→ In each of the following squares there are basic lines and geometric objects. Create a drawing in each of the 6 squares, using all the lines already provided.

For example, you could do this...

→ Repeat the exercise as many times as you wish, either by using the lines above, or creating new starting points for yourself or your friends and family.

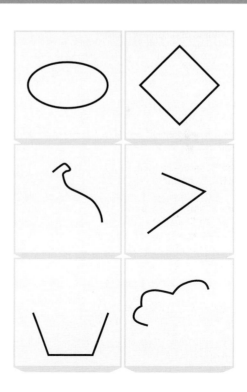

Turn to pages 144 to 155 for the answers.

Exercises to improve the left brain

FOR THOSE WITH A RIGHT-BRAIN BIAS

VERBAL EXERCISES

1. **Solve the following one-word anagram**

   SEND RAIL

2. **Which is the odd one out?**

   (a) wolverine
   (b) tiger
   (c) tiercel
   (d) leveret
   (e) camel

3. **What is a KNOUT?**

   (a) canal
   (b) bridge
   (c) whip
   (d) raft
   (e) castle

4. **Solve the one-word anagram**

   LOUDER NAY

5. **Which is the odd one out?**

   salsa, samba, foxtrot, pavid, charleston, galliard, waltz?

6. **What is the name given to a group of mallards?**

   (a) intelligence
   (b) parliament
   (c) sord
   (d) medley
   (e) flush

7. **Make a seven-letter word from the following five letters**

   EGUSA

8. **Place three-letter clusters together to make a six-letter bird**

   swa, gin, gee, can, row, ary, spa, ser, pen, loy

9. **Make a six-letter word from these four letters**

   NELK

10. **Which is the odd one out?**

    (a) corduroy
    (b) marcasite
    (c) damask
    (d) taffeta
    (e) linen

→ Try to interpret each of the 20 drawings below in the wildest possible way.

For example, you may think at first glance that the first drawing is a crescent moon shining brightly out of the dark night sky. But couldn't it also be the glint from the eye of a one-eyed cat peering at you from a pitch-black cellar?

If you wish to measure your creativeness, try the exercise in the company of other people. The main thing is to let your imagination run wild and lose your inhibitions. The more people laugh at your efforts, the more successful you are likely to have been in using your imagination.

HAVE FUN!

Exercises to improve the left brain

FOR THOSE WITH A RIGHT-BRAIN BIAS

TECHNICAL EXERCISES

1. **Of the nine planets, which is the smallest?**

2. **What does this mathematical diagram mean?**

3. **What is the name of the line A to B?**

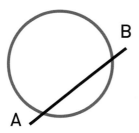

4. **What is the possibility of scoring at least 10 with two standard dice?**

5. **What is the surface area of a ball?**

6. **If you wanted to make a square-based water tank out of sheet metal without a lid, how would you make it with the minimum amount of metal?**

7. **What is a curve called on which a ball is placed anywhere on the inner surface that will always take the same time to arrive at point A?**

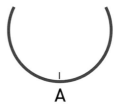

8. **What is the size of a standard house brick in imperial and metric, to the nearest quarter inch, and the nearest 5mm? How many bricks would be needed for a half brick wall 9 inches wide and 6 inches high?**

9. **How many sides do snowflakes have?**

SPATIAL AWARENESS: TANGRAMS

→ Tangrams are an ideal way to exercise your creativity, spatial awareness, imagination, and artistic talents.

The tangram puzzle is probably the most ancient in origin of all dissection puzzles. It is believed to have originated in China 4,000 years ago, although the earliest known reference to it is a woodcut from 1780 by Utamoro, which depicts two courtesans trying to solve Chi-Chiao (the seven clever pieces).

The puzzle is made up of seven pieces cut from a square as illustrated below, and the object is to create tangram shapes from the seven pieces.

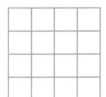 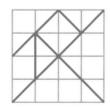

It is very easy to make a set of tangram pieces yourself:

Cut out a square piece of card; about 3 in (8 cm) square is a good size. Lightly mark a 4 x 4 in (10.5 cm) grid of squares on it (see above), for the purpose of guidelines. Next, draw in the lines shown above left in green and cut along the lines. The result is the seven tangram pieces shown below.

These seven pieces are the basis of all tangram puzzles and any shapes devised from them must use all seven pieces, with no pieces overlapping.

The author of *Alice in Wonderland*, Lewis Carroll, himself a great compiler of puzzles, was fascinated by tangrams and owned a book, *The Fashionable Chinese Puzzle*, that contained 323 tangrams. On his death, it passed to another puzzle inventor, H.E. Dudeney, who created the tangram below of *The Mad Hatter*, a character from *Alice in Wonderland*.

Other examples of tangram figures, a yacht, a cat, and a native American are shown below.

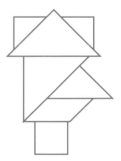

All seven figures are used, and no pieces overlap.

Continued on page 143

Exercises to improve the left brain

FOR THOSE WITH A RIGHT-BRAIN BIAS

TECHNICAL PUZZLES

1. A man has a stack of 27 square blocks. Somebody painted the blocks without disturbing them in the night.

   How many blocks had paint on them?

2. A decimal number is 119.

   What is it in binary?

3. What is meant by the mathematical sign below?

4. What is the volume of a cylinder?

5. Lengths of pipe are to be packed in hexagonal bundles. The pipes that can be seen on the outside total 18.

   How many are in the bundle?

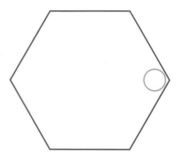

6. Brass is an alloy made up of two metals.

   What are they?

7. There are only five polyhedra, i.e. solid regular figures.

   Which one has 20 faces?

   (a) icosahedron
   (b) dodecahedron
   (c) cube
   (d) tetrahedron
   (e) octahedron

8. This is a weather symbol. What does it mean?

Continued from page 141

SPATIAL AWARENESS:

→ H.E. Dudeney created the following classic tangram paradox known as *The Two Chinamen*.

The two men are apparently identical, except for the missing foot on the right-hand figure. Yet, surprisingly, both figures contain all seven identical pieces. How can this be so?

It is time to put your creative talents to the test and try to figure out the solution.

Answer:

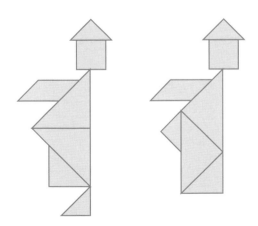

→ Below is another set of nine figures. It is possible to construct the figures with the seven tangram pieces.

→When you have solved these, continue to devise new figures and geometric shapes. You may be surprised at the number of different shapes and figures it is possible to create.

Answers to Section Two

MATHEMATICAL CALCULATION: CONNECTIONS
(exercises on pages 74, 76, and 78)

1.

2.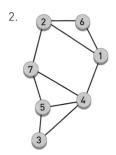

3.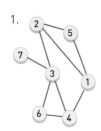

4.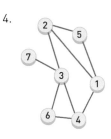

5.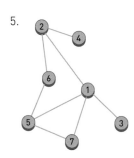

6.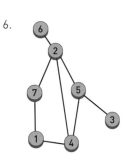

7.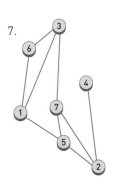

8.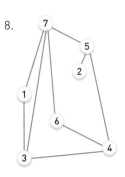

9.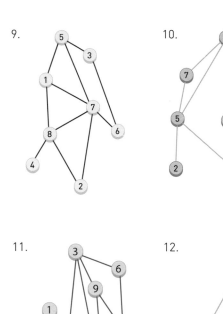

10.

11.

12.

LATERAL NUMBER PUZZLES
(exercises on pages 77, 79, 81, and 83)

1. **12.** The number at the bottom is formed by deleting all the numbers that appear twice in the adjoining triangles and adding the numbers remaining.

2. **78.** Opposite circles connected by just one line contain successive digits (e.g. 3456). Opposite circles connected by two lines contain alternate digits (e.g. 2468).

3. **14.** Looking across and down each line of numbers, the number in the first small circle is the sum of the first digits of the numbers in the large circle, the number in the second circle is the sum of the second digits.

4. **2.** Numbers in the same position in each grid total 10.

5. **19.** Looking at the small ovals clockwise, the odd numbers appear in the sequence 79135791. Looking at the large ovals clockwise, the even numbers appear in the sequence 208642.

Answers to Section Two

6. **138.** Each number is determined by the two numbers immediately below it. In rows 2 and 4 the rule is the sum of two numbers below plus 1, and in rows 3 and 5 it is numbers below minus 1.

7. **1** in both cases. When the same figures are adjacent, the number doubles if the figure increases in size and halves when it decreases in size. It stays the same when it remains the same size. If different figures are adjacent, the number increases by 1 if the figure next to it has more sides, and decreases by 1 if it has less sides.

8. **18.** The figure 1924 is determined by taking the other 4 digits as follows: 1^2, 3^2, $4 \div 2$, $8 \div 2$.

9. **9.** Each number is totaled by counting the number of sides in the adjoining two figures.

10. **4.** The numbers 36942 are repeated in the pattern shown at the top left of the next column.

11. **9.** The opposite numbers, as shown in the diagram at the top right of the next column, are the sum of digits of their connected number.

12. **5.** The number in the middle is the sum of the numbers on either side. Where there are four numbers in the row, the number formed by the middle two squares is the sum of the two numbers on either side.

13. **1.** The four cubes are the visible faces of a die being rolled south, west, north, and east.

14. **7.** The totals in each column decrease in the series—22, 20, 18, 16, 14.

15. **5.** Looking around the figures clockwise in the pattern shown in the next column, the rule is: hexagon with circle add 4, pentagon with diamond deduct 5, pentagon with circle deduct 4, hexagon with diamond add 6.

16. **82.** Looking at the bottom square in conjunction with the two squares above, numbers in the same corners total 100.

17. **1410.** Looking northwest to southeast, add the odd digits of the two outside circles to obtain the first part of the number in the middle. Looking northeast to southwest, add the even digits to obtain the second part of the number in the middle.

18. **15.** Each of the numbers in the pentagons is the sum of two of the digits in the hexagon. There are 15 possible pairings of the six numbers.

19. **0.** Looking at diagonal lines from bottom to top starting with 4, the numbers formed are $4 \times 4 = 16 \times 8 = 128 \times 16 = 2048 \times 32 = 65536$.

20. **2.** On both the top row and bottom row, the sum of numbers in upright triangles is 16, and so is the sum of numbers in inverted triangles.

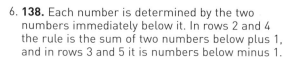

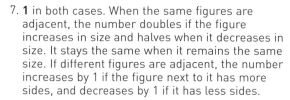

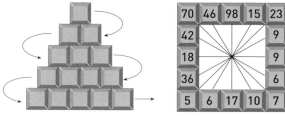

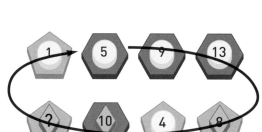

DEFINITIONS
(exercises on pages 80 and 82)

Definitions Test 1

female demon	lamia
relating to animal fat	adipose
fleshy part of plants	herbage
small, delicate, and charming object	bijou
over-fussy attention to details	punctilio
fine parchment	vellum
warning of danger	monition
encouraging, urging on	hortatory
heroic knight or hero	paladin
bland or uncontroversial	anodyne

Definitions Test 2

to turn aside	distract
military store	ordnance
abiding in, inherent	immanent
show off	flaunt
subject to persistent ill-treatment	persecute
bring legal proceedings against	prosecute
decree	ordinance
to take away from	detract
show contempt for	flout
impending, close at hand	imminent

Definitions Test 3

cleverly contrived	ingenious
standing still	stationary
in that place	there
belonging to them	their
writing materials	stationery
impartial	judicial
unwilling	loath
prudent	judicious
naive, innocent	ingenuous
dislike intensely	loathe

Definitions Test 4

supplies, especially military	materiel
servile, ingratiating	obsequious
chief	principal
one who hates women	misogynist
priority	precedence
substance of which something is made	material
to do with funeral rights	obsequial
examples	precedents
one who hates marriage	misogamist
fundamental truth	principle

MATHEMATICAL CALCULATION
(exercises on pages 84 and 86)

1. **3 minutes fast.**

Slow		**Fast**	
My watch	15 minutes	Stove clock	25 minutes
Clock	5 minutes	Church clock	30 minutes
Your watch	20 minutes		

   The difference is 15 minutes fast.
   15 divided by 5 = 3.

2. $18

3. **2 hours.** 40 mph for 80 miles = 2 hours;
   60 mph for 2 hours = 120 miles.

4. **30 mph.** 210 divided by 30 = 7;
   210 divided by 35 = 6.

5. **6 minutes.**
   5 x 4 x 3 x 2 x 1 = 180 x 2 = 360 seconds,
   or 6 minutes.

6. **He has $2.40 left.** Three-fifths of 420 = 252,
   leaving a balance of 168 (420-252);
   0.45 x 168 = 75.6, leaving a balance of 92.4
   (168 - 75.6); 92.4 - 90 = 2.40.

7. 24 minutes before 12:00 P.M., or 11:36 A.M.
   11:36-84 minutes = 10:12 A.M.
   or 72 minutes (3 x 24) past 9:00 A.M.

8. **Tom $21,000, Dick $42,000, Harry $77,000.**

15,000	Tom	15%	21,000
30,000	Dick	30%	42,000
55,000	Harry	55%	77,000
100,000		100%	140,000

9. Tony 175, Sally 140, Pat 112.

10. **Three times as old.**

	Father	Son
x 2	32	16
x 5	20	4
x 4	24	8

11. 24

12. **J = $7.50   H = $2.50**
    Each man should have 4 loaves after the bread is
    shared. John has 7 loaves, so he gives 3 loaves
    to Charlie. Henry has 5 loaves, so he gives 1 loaf to
    Charlie. John therefore receives 3 times as much
    money from Charlie as Henry.

13. A: $126 (42 % of $300)
    B: $111 (37 % of $300)
    C: $ 63 (21 % of $300)

14. 6 x 5 x 4 x 3 x 2 x 1 = **720**

15. Man $\frac{1}{80}$ = .0125

    Wife $\frac{1}{200}$ = $\underline{.0050}$

    $.0175 = \frac{1}{0.175}$ = **57.14 days**

16. $\frac{9 \times 8 \times 7 \times 6 \times 5}{1 \times 2 \times 3 \times 4 \times 5}$ = **126**

17. **$188**

C:	$50	= 50
B:	$50 + 20 percent	= 60
G:	50 percent x $110 (B+C)	= 55
F:	20 percent x $115 (B+G)	= $\underline{23}$
		$188

18. **1.8 hours**
    A: 4 hours = $\frac{1}{4}$ = 0.25

    B: 5 hours = $\frac{1}{5}$ = 0.20

    C: 10 hours = 1 = $\frac{0.10}{0.55}$ $\frac{1}{0.55}$ = 1.8 hours

19. **Barbara is 37 years old.** 37 x 2 = 74
    74 minus 1 = 73. Reverse 73 = 37.

20. **10 chances in 216.**
    10 ways to score 15 with 3 dice
    Can score 216 with 3 dice (6 x 6 x 6)

21. $\frac{4}{52} \times \frac{3}{51} \times \frac{2}{50} \times \frac{1}{49} = \frac{1}{270725}$

22. 49 times in 50
23. One way is 123 − 4 − 5 − 6 − 7 + 8 − 9 = **100**
24. 50 × 101 = **5050**
25. **184,926.** Since 37 is one-third of 111, each fish has the equivalent of 37 spots.
26. $6^4 ÷ 48 = 27$
27. Bill $180, Alan $140
28. $(7-5)^2 + 96 + 8 − 4 − 3 − 1 = 100$
29. A: 4 in 7; B: 2 in 7; C: 1 in 7
30. Each letter of the alphabet is allocated a number.
So, W = 8, E+ 1, and so on.
    81729
    678912
    <u>217669</u>
    978310

LETTER AND WORD PUZZLES
(exercises on pages 85 and 87)

1. **Print.** Each word can be prefixed with the colors of the rainbow in turn: red squirrel, orange squash, yellow streak, greenhouse, blueprint.
2. **E.** Convert each letter to its numerical letter in the alphabet. Each vertical and horizontal line totals 27.
3. They all contain two consecutive double letters.
4. **K.** Start at Z and work backwards through the alphabet in the sequence:
ZY(x)W(vu)T(srq)P(onml)K(jihgf)E
working clockwise and jumping two segments each time.
5. **CGK.** Each oval starts with consecutive letters of the alphabet ABCD. Each oval jumps one letter, two letters, three letters, four letters of the alphabet in turn. So, **C**def**G**hij**K**.
6. **X.** Take one letter from each cube in turn to spell out ONE, TWO, SIX.
7. **lap and outplayed.** Each of the other three letter words is spelled out backwards in the middle of one of the nine-letter words: arrogance/ago, fantastic/sat, sentenced/net, elaborate/rob.
8. **N.** Start with A, which is formed with three lines, then take the next letters you come to in the alphabet that are formed with four lines followed by three lines alternately.
9. **Q.** Each group is of a string of alternate letters of the alphabet; OpQrStU.

VERBAL INTELLIGENCE
(exercises on pages 88 and 90)

1. safe
2. ineffective
3. fantasy, delusion
4. wreathe, envelop
5. [d] persevering in one's duty
6. mystic, arcane
7. pithy, wordy
8. [d] carefree and lighthearted
9. innocuous
10. blame
11. abandon, eschew
12. inhibit
13. dominate, ancillary
14. [a] severe criticism
15. activate, arrest
16. [e] containing iron salts
17. capacious
18. barren
19. illusion, chimera
20. singular, conventional
21. [a] the number 10
22. basinet (a soldier's headpiece)
23. head
24. dictionary
25. abundance
26. [c] exultation
27. wan, florid
28. filly
29. [c] a sundial
30. tympanum

ODD ONE OUT VISUAL EXERCISES
(exercises on pages 89 and 91)

1. **D.** In all the others, the orange spirals are curved and the green spirals have straight lines.
2. **E.** The rest represent something with a color in its name: bluebell, red star, red square, blue moon.
3. **C.** All the others have three identical figures when rotated. C contains a figure that is a mirror image of the other two, not a rotation.
4. **E.** In all the other four-sided figures there are two red dots and one blue dot, and five-sided figures have two blue dots and one red dot.
5. **B.** It is on its own because it is a green pointer at a blue star. All the others are in pairs: A and C are red pointers at blue stars; D and E are blue pointers at red stars.
6. **A.** All the others have the same sequence of little ovals clockwise. A has the same sequence, but counterclockwise.
7. **D.** A and E are the same with red/green reversed, like B and C.
8. **D.** In all the others, the color of the bottom ring is repeated at the top.
9. **B.** It is the only one where two blue dots are directly connected.
10. **B.** The total number of sides in the two figures is 9. In the others it is 10.

Answers to Section Two

LETTER/NUMBER LOGIC
(exercise on page 92)

• •

22. In column 4, line 3. Letters and numbers are mirrored symmetrically in the top and bottom halves of the grid, like black squares in a crossword grid. Letters in the top half are mirrored in the bottom half by their numbered position from the beginning of the alphabet. Letters in the bottom half are mirrored in the top half by their numbered position from the end of the alphabet.

ANALYTICAL PUZZLE
(exercises on pages 92 and 94)

• •

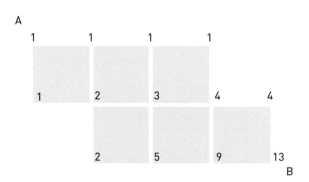

1. **13 ways.** Start at A and mark it with a 1, because from that point there is only one way to go. Put a 1 at each corner where there is only one way to get there from A. Put a 2 at a corner where it is possible to arrive in two different ways. By continuing in this way, it can be seen that the number in each corner is the sum of the numbers along the paths leading to that corner.

2. **12 inches.**

   Based on Pythagorean theorem,
   x = diameter 13 inches
   y = 12 inches
   z = 5 inches
   $5 \times 5 = 25$
   $12 \times 12 = \underline{144}$
   $13 \times 13 = 169$

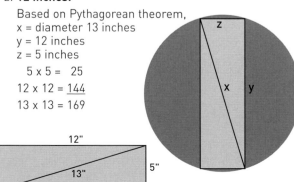

3. 56
4.
$$1 = 16 = 4^2$$
$$2 = 9 = 3^2$$
$$3 = 4 = 2^2$$
$$4 = \underline{1} = 1^2$$
$$30 \quad 30$$
5. 25

CREATIVE SEQUENCE EXERCISES
(exercises on pages 93 and 95)

• •

1. There are two sequences. In the vertical diamonds, the blue dot moves up and down. In the horizontal diamonds, the red dot moves up and down.

2. The circles are constructed a segment at a time, moving clockwise.

3. There are three shapes of triangles repeated, and two colors, yellow and purple, repeated.

4. A blue, then a red, dot is added vertically and horizontally at each stage. Only the combination that appeared at the previous stage is linked.

5. The figure tumbles through 45 degrees at each stage, and the yellow portion moves to a different position at each stage, moving clockwise.

6. The orange rectangle moves from right to left one place at each stage.

7. At each stage the yellow circle moves between two opposite corners, the purple dot moves one place counterclockwise, and the black dot moves one place counterclockwise.

8. Divide the circles into groups of three. The blue circle moves down one place at each stage, for example,
   blue/red/yellow
   red/blue/yellow
   red/yellow/blue

9. Moving clockwise, a different corner is turned over each time, alternating red and yellow.

10. At each stage the blue dot originally at the three o'clock position moves one place counterclockwise, changing places with the dot immediately to the left of it.

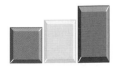

148

MATRIX PUZZLES
(exercises on pages 96 and 98)

1. **B.** Looking across and down, the figure tilts 45 degrees counterclockwise, and each row and column contains one black figure.

2. **F.** Looking across and down, the third square is the first two squares combined.

3. **C.** Looking across and down, the third square is combination of the first two squares, except that lines that appear in the same position in both the first two squares are not carried forward to the third square.

4. **A.** Looking across and down, the third square is a combination of the first two squares, except that lines that appear in the same position in both the first two squares are not carried forward to the third square.

5. **F.** Only elements that are common to the first two squares are carried forward to the third square; however, rectangles turn to circles, and vice versa.

6. **A.** Looking across and down, only lines that appear in the same position in the first two squares are carried forward to the third square.

7. **F.** Only elements that are common to the first two squares are carried forward to the third square; however, turquoise turns to brown, and vice versa.

8. **B.** Looking across and down, the number of circles in the final square is the difference of the totals of circles in the first two squares, so if the first two squares contain four red and two blue circles, two red circles appear in the final square.

9. **D.** Looking across, add up the number of each colored circle and put that number of circles in the final square; however, change around the colors. Looking down, simply add up the number of each color in the first two squares and put that number in the bottom square.

10. **B.** Looking across and down, the yellow arc moves 90 degrees counterclockwise at each stage, and the red arc moves 90 degrees.

SPATIAL AWARENESS
(exercises on pages 97, 99, 101, 103, 105)

1. **B.** Everything previously on the inside goes on the outside, also red triangles change to blue circles, and vice versa.

2. **D.** Colors change sides as in the original analogy.

3. **B.** Each section is a mirror image of the section opposite, but with the colors reversed.

4. **A.** The shapes and colors change places as in the first analogy.

5. **D.** Each circle has a pairing that has turned 90 degrees.

6. **B.** The large figure rotates 180 degrees clockwise, and the smaller figure moves to the top of it.

7. **D.** In the other figures, there are twice as many branches on the right side as on the left side.

8. **C.** It is the only one where in the string of two green and one yellow beads, the yellow bead is in the middle.

9. **A.** It is the only one that is not a mirror image of the squares above and below it.

10. **C.** In the bottom circle, all the small circles are a mirror image of one of the circles in the top set.

11. **D.** B and E are the same with blue/green reversal. A and C are the same with blue/red reversal.

12. **C.** Red lines become green and blue lines become dotted red (see below).

13. **C.** The others are the same figure rotated.

14. **C.** Colors are carried forward when they appear twice in the same position in the first three rectangles.

15. **D.** Dots change color as in the first analogy, so brown becomes dark blue, green becomes light blue, and red becomes yellow.

16. **9.** All the rest have a mirror-image pairing with colors reversed.

17. **E.** Each horizontal and vertical line contains one each of the three different patterns.

18. **A.** The two halves of the figure simply change around.

19. **B.** The section common to the triangle and square should be green, not red.

20. **A.** In all the others, the color on the outside is repeated in the middle.

21. **G.** In all the others, there are five red and five blue sides in total.

22. **A.** The sequence appears by looking at dots in the same position in each pentagon. Starting at the top and working clockwise, the repeated sequences are; green/orange, orange/green/blue, blue/orange, green/blue, and blue/orange/green.

23. **D.** B reverses the sequence of C, and A reverses the sequence of E.

Answers to Section Two

MAGIC SQUARES
(exercises on pages 100, 102, and 104)

1.

16	3	2	13
5	10	11	8
9	6	7	12
4	15	14	7

2.

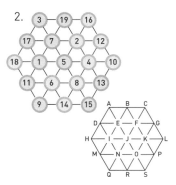

3.

14	3	11	13	24
19	23	7	10	6
20	15	1	17	12
4	22	25	9	5
8	2	21	16	18

25	10	3	6	21
22	12	19	8	4
11	9	13	17	15
2	18	7	14	24
5	16	23	20	1

4.

17	5	10	20	13
16	23	14	8	4
11	7	1	25	21
2	24	18	9	12
19	6	22	3	15

5.

6	7	19	18	25	36
32	11	14	20	29	5
33	28	16	22	9	3
4	27	15	21	10	34
35	8	23	17	26	2
1	30	24	13	12	31

6	7	19	18	25	36
32	11	14	20	29	5
33	28	16	22	9	3
4	27	15	21	10	34
35	8	23	17	26	2
1	30	24	13	12	31

6.

24	19	26	6	1	35
25	23	21	7	32	3
20	27	22	2	9	31
15	10	17	33	28	8
16	14	12	34	5	30
11	18	13	29	36	4

7.

15	2	9	18	21
8	16	25	12	4
22	14	3	6	20
1	10	17	24	13
19	23	11	5	7

WORD CHANGES
(exercises on page 106)

1. Education is the **method** whereby one acquires a higher **grade** of prejudices.

2. Tsunamis start out as barely noticeable deepwater **ripples** caused by underwater earthquakes or volcanic **eruptions**.

3. An effective business telephone style can only be achieved by **constantly** reviewing the practices we **personally** develop as time goes by.

4. Prior to any **interview**, the interviewer must have a clear picture of what he or she is looking for in the candidate being interviewed in relation to the requirements of the **job**.

5. Today, **people** are drinking the wines they enjoy, irrespective of whether the so-called wine **buffs** still embrace the old-fashioned **values**.

6. Early **morning** is usually set aside to accomplish set **routines**; late **afternoon** appointments make it more difficult to complete the day's work.

7. Libraries have comprehensive **index** facilities at their disposal, and even if the **information** is not available from their **reference** material, they will know who you should contact.

8. From the **tee** high above the **lake**, the distance across to the **fairway** looks shorter than it is, as distances invariably do over water.

9. Truly **awful** writing can transcend its own **inadequacies** to rise to **surreal heights**.

10. The **spread** of **elementary education** in the late nineteenth century spawned a **generation** of tenth-rate **popular novelists**.

CREATIVE SOLUTIONS
(exercise on page 107)

It should be possible to attach the candle to the door with the thumbtacks, or alternatively melt some wax and stick the candle to the door. However, by far the most efficient way is to empty all the thumbtacks out of the box and pin the box to the door with the thumbtacks. Then stick the candle in the box and light it. This way, no wax drops on the floor, the candle should not drop out of the box, and it can be adjusted to give out light in the required direction.

MATHEMATICAL CALCULATIONS: BRAIN STRAIN
(exercises on pages 108 and 110)

1.
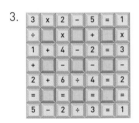

8	x	2	–	9	=	7
+		x		+		–
8	–	6	x	2	=	4
÷		–		–		x
4	+	8	÷	6	=	2
=		=		=		=
4	÷	4	+	5	=	6

2.
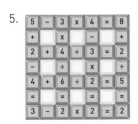

5	x	4	÷	5	=	=
+		+		–		+
7	–	6	+	3	=	4
÷		–		x		–
6	+	8	÷	2	=	7
=		=		=		=
2	+	2	÷	4	=	1

3.
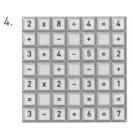

3	x	2	–	5	=	1
÷		x		+		x
1	+	4	–	2	=	3
+		–		–		–
2	+	6	÷	4	=	2
=		=		=		=
5	–	2	÷	3	=	1

4.
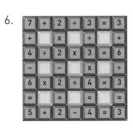

2	x	8	÷	4	=	4
+		–		+		+
3	+	4	–	5	=	2
–		÷		–		–
2	x	2	–	3	=	1
=		=		=		=
3	–	2	+	6	=	7

5.

5	–	3	x	4	=	8
+		x		–		+
2	+	4	÷	3	=	2
–		÷		x		÷
4	+	6	÷	2	=	5
=		=		=		=
3	–	2	x	2	=	2

6.

7	+	2	÷	3	=	3
+		x		÷		+
4	÷	2	x	3	=	6
–				x		÷
6	x	2	÷	4	=	3
=		=		=		=
5	+	2	–	4	=	3

POOL AND SNOOKER
(exercise on page 109)

1.

2.

3.

CREATIVE LOGIC
(exercises on pages 111, 113, 115, 117)

1. **C.** The blue triangle remains stationary while the green triangle moves clockwise around the blue triangle, pivoting on its leading edge.
2. **B.** When two dots of the same color appear in a circle, they disappear and are replaced by two different-colored dots at the next stage.
3. **A.** Looking across and down, the contents of the third square are determined by the contents of the first two squares. Only when the same color appears in the same position is a transfer made to the third square; however, red turns to yellow, and vice versa.
4. **D.** Reverse the color sequence of the previous figure and discard the color next to the end.
5. **E.** The red bar moves left to right one place at each stage, and the purple bar moves right to left. The yellow and green bars always occupy the two spaces not occupied by the purple and red bars.
6. **B.** Start at the bottom left square and work up the first column, then around the perimeter and gradually spiraling into the center, repeating the colors green, yellow, pink, blue.
7. **C.** The yellow/blue pentagon moves counterclockwise around the green/yellow pentagon a side at a time. In the green/yellow pentagon, the yellow portion moves around clockwise at each stage.
8. **C.** Work clockwise, changing orange to blue, first missing no orange lines, then one orange line, then two orange lines, and so on.
9. **C.** A square is being formed by adding an orange then blue line alternately at the top left, and pushing the lines added already around counterclockwise.
10. **E.** There are three shapes of triangles and two colors, red/yellow.
11. **D.** Purple and green alternate, the purple tubes become smaller, and the pink larger.
12. **A.** The blue line moves one side counterclockwise, and the green line one side clockwise at each stage.
13. **B.** All lines are carried forward to the third square, across and down; however, where two red lines appear in the same position in the first two squares they change to blue, and vice versa.
14. **D.** Looking across the line of squares, the sequence is: top left corner, red/blue/green; top right corner, black/pink; bottom left corner, yellow/purple/red; bottom right corner, turquoise/brown.
15. **E.** Starting at the east arm and working clockwise, two circles change places in turn: green/blue, blue/red, red/red.
16. **B.** Every pair of suits changes places in turn. Spades/diamonds is the final pair to change places.

MONOGRAMS
(exercises on pages 112, 114, and 116)

1. Leonardo (da Vinci)
2. Arc de Triomphe
3. Pythagoras
4. Santa Claus
5. Roosevelt (Franklin D.)
6. Churchill (Winston)
7. Garibaldi
8. Beethoven
9. Tour Eiffel (Tower)
10. Columbus

WORDS
(exercises on page 118)

1.

1. Gig
2. Moor
3. Desert
4. Flag
5. Char
6. Blubber
7. Kid
8. Peal
9. Pinch
10. Dub

2.

1. (d) pod
2. (b) unkindness
3. (a) herd
4. (d) observation
5. (a) knot
6. (c) murmuration
7. (c) state
8. (a) paddling

HEXAGONS
(exercises on pages 119 and 121)

The rule is: 1 is added to 2 to make 3,
4 is added to 5 to make 6,
but similar symbols disappear.

1. B
2. B
3. E
4. C
5. B
6. B
7. D
8. E
9. C
10. A

FORE WORDS
(exercise on page 120)

1. Pin
2. Foot
3. Hard
4. Cat
5. Corn
6. Gold
7. Man
8. Long
9. Car
10. Draw
11. War
12. God
13. Can
14. Star
15. Loud
16. Sun
17. Main
18. Tap
19. Book
20. Hold

TECHNICAL
(exercises on pages 122, 124, 126, 128, and 130)

1.

H Γ Δ Θ Υ

Eta   Gamma   Delta   Theta   Upsilon

2. 3^2 x 16 feet = 144 feet. The formula is: height = time squared x 16 feet

3. $\dfrac{4}{3} \pi R_3$

4.

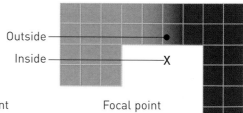

Focal point
Area ½ distance from focal point

12 x 1½	=	18
6 x 4½	=	27
12 x 7	=	84

30 divided into 129
= 4½

Turn 90 degrees

12 x 2	=	24
6 x 1	=	6
12 x 3	=	36

30 divided into 66
= 2⅕

Focal point
Center of gravity

5. Force 9, according to the Beaufort scale of wind speeds.

6. (d) 61.8 percent Actual size 0.618034 of the large plate

7. ESE

8. (a) plus or minus (3)
(b) is greater than (5)
(c) is not equal to (11)
(d) square root of (9)
(e) infinity (7)

9. **C to F**: C x 9/5 + 32 degrees
   F to C: F -32 degrees x 5/9

   -40 degrees centigrade
   -40 degrees fahrenheit
10. (**b**) 33 degrees
11. (**b**) 1½" x 5"

    The formula is BD3
    B = breadth   D = depth

    So:

2" x 4" = 32"	(4 x 4 x 2)
1½" x 5" = 37½"	(5 x 5 x 1½)
3" x 3" = 27"	(3 x 3 x 3)
1" x 6" = 36"	(6 x 6 x 1)
1¾" x 4½" = 35 7/16"	(4½ x 4½ x 1¾)

12. (**d**) parabola
13. (**c**) 316 cubic inches

    ⅓ height x circle (πr^2)

    $4 \times 3\frac{1}{7} \times 5^2 \left(\frac{10}{2}\right)$ = 316 cubic inches
14. **30 to 6 or 5 to 1**

    1st Die   1 2 3 4 5 6

    2nd Die   1 1 1 1 1 ~~1~~
    2 2 2 2 ~~2~~ 2
    3 3 3 ~~3~~ 3 3
    4 4 ~~4~~ 4 4 4
    5 ~~5~~ 5 5 5 5
    ~~6~~ 6 6 6 6 6

15. 820 feet
16. **1.35 hours**

    Take reciprocals (e.g. a reciprocal of 2 = $\frac{1}{2}$ = 0.5)

    3 = 0.333
    6 = 0.166
    2 = 0.500
    10 = 0.100
    4 = 0.250
    1.350

    Take reciprocal 0.73 hours

    $\frac{1.000}{1.350}$
17. This breaks down as follows:

1	1	1	1	1	.	1	1	1	
16	8	4	2	1	.	½	¼	⅛	= **31⅞**

18. $\frac{360}{5}$ degrees = 72 degrees

    180 degrees - 72 degrees = **108 degrees**
19. **32 kg.** 20 kgs + ($\frac{3}{8}$ x 32 kg)
20. Saturn

Colors, Circles, Squares, Pentagons
(exercises on pages 123 and 125)

1. **D.** Pink moves clockwise one angle; green moves counterclockwise two angles; blue moves clockwise two angles
2. **B.** A-B-C repeats, so D is the same as A; E will be the same as B; F will be the same as C
3. **E.**

	B P G
A	2 1 1
B	1 2 1
C	2 2 -
D	1 2 1
E	2 1 1

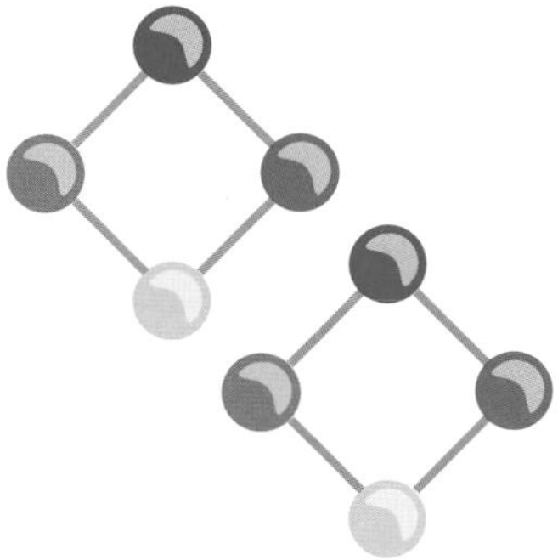

4. YGPB 5/2/3/10

5. Blue. YGPB YGPB YGPB

6. **J.** A–F B–E C–K D–G H–I

7. **D.** In each row, the left square is added to the middle square. Identical symbols are eliminated to make up the third square.
8. **B.** Two blues change to two pinks, two greens change to two blues, two pinks change to two greens
9. **D.** Order should go G P B P B B clockwise, this figure goes G P B P B
10. **B.**
    Green moves two angles counterclockwise
    Blue moves two angles counterclockwise
    Pink changes to green, to blue, to pink, to green, and so on
    Yellow moves 3 angles clockwise

Symbols, Grids, and Others
(exercises on pages 127 and 129)

1. **E**
2. **2B.** One dot is missing.
3. **D.** A is the same as E; B is the same as G; C is the same as F.
4. **E**
5. **D.** Each circle is made by joining up the symbols in the two circles below, eliminating identical symbols.
6. **2B.** One dot is missing.

CREATIVE NUMBER PUZZLES
(exercises on page 131)

1. **21322314.** This is a repeat of the number above. The sequence carries on repeating this same number over and over again. Each number describes the number above, by sorting first the 1s, then 2s, then 3s, then 4s. So the number 21322314 has two 1s, three 2s, two 3s and one 4.

2. **49.** It is the position of each letter *e* in the question.

ALTERNATING PUZZLE
(exercise on page 131)

You need to pick up only two glasses. Pick up 9 and pour the contents into 4, and pick up 7 and pour the contents into 2.

LOGIC
(exercises on page 132)

1. **[d].** $4x^2 - 4xy + y^2$
2. 5⅓kg

   7kg x 3 = 21   4kg x 2 = 8

   $$5\tfrac{1}{3}kg \times 6 = \frac{32}{53}$$ $$9kg \times 5 = \frac{45}{53}$$

3.
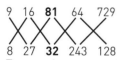

   9   16   **81**   64   729

   8   27   **32**   243   128

   Top numbers are multiplied by 3 and the answer goes to the bottom row (e.g. 9 x 3 = 27).

   Bottom numbers are multiplied by 2 and the answer goes to the top row (e.g. 8 x 2 - 16).
4. **454** — 39, 40, 41, 42, 43, 44, 45, 46
5. $\frac{7}{41} \times \frac{123}{28} = \frac{3}{4}$
6. **10½**

   There are two series   (+3½) 4½, 8, 11½, 15

   (-7½) 33, 25½, 18, 10½
7. **22**

   Modulus 10   Modulus 9

   7x6=42   7x6=46

   4x5=20   4x5=22

8. Add 6 – (40) + 4 -12= **- 42**
9. $$\frac{6}{1\tfrac{2}{3} - \tfrac{1}{2}} = \frac{6}{1\tfrac{1}{4} - \tfrac{3}{8}} = \frac{6}{1\tfrac{1}{6}}$$

   $$= \frac{6 \div 7}{6} = \frac{6 \times 6}{7} = \frac{36}{7} = \mathbf{5\tfrac{1}{7}}$$

10. **[d]** mandrill
11. salmon
12. **[c]** goblin
13. **[c]** doldrums
14. **[b]** rgwei
15. **[c]** discerning
16. **[d]** defilade
17. head
18. tours, azure

CREATIVITY: THE TRICK DONKEYS
(exercise on page 133)

MATHEMATICAL LOGIC
(exercises on page 134)

1. **12.** There are two series:

   6, 7½, 9, 10½, 12 (+1½)

   23, 21, 19, 17 (-2)
2. **+17**
3. 6 – (34) – (7) = **-35**

   (Order must be taken ( ), x, ÷, +, -)
4. $\frac{64 \times 5}{10 \quad 8} = \frac{8}{2} = \mathbf{4}$
5. **81 degrees**
6. 27; 96 – 69 (reversed 96) = **27**
7. **107** (54 + 53)
8. **–15½** (-8⅛). Less 8⅛ at each step. 17, 8⅞, ¾, -7⅜, -15½
9. **32** (6 x 6) – (2 x 2)
10. (3 + 2) - (16² - 9²)

    5 - 175 = **-170**

MATHEMATICAL CALCULATION
(exercises on page 136)

1. **The total time is 60 seconds.**
   Call the slices A, B, and C. Fry A and B on one side. Swap B for C and fry A on the other side, and C on its first side. Now swap A for B, and fry the second side of B and C at the same time.

2. **The order makes no difference.**
   A $100 suit would be $64 if the discounts were taken in any order.

3. **66 years.** 66 + 33 (½ of 66) + 22 (⅓ of 66) + 9 (3 x 3) = 130 = 6 score + 10

4. **40 miles per hour.**
   30 miles in 60 mins + 30 miles in 30 mins = 60 miles in 90 mins (which is 40 miles per hour).

5. A $210   B $122.50   C $17.50

6. **174 heads and 582 legs.**
   Heads: 117 quadrupeds + 57 birds = 174 heads.
   Legs: 468 quadrupeds + 114 birds = 582 legs.

7. **60 days** (3 x 4 x 5)

8. 9, 7, 4, 11, 14. Omit 9 = 36; omit 7 = 38; omit 4 = 41; omit 11 = 34. omit 14 = 31.

9. 2½ lbs. $\frac{5}{7}$ lbs + ($\frac{5}{7}$ x 2$\frac{1}{2}$) = 2$\frac{1}{2}$

10. 70 + 75 + 80 + 85 = 310
    - (3 x 100)   300 -
                    10%

11. Letters of the alphabet

12. $\frac{17 \times 16 \times 15 \times 14 \times 13 \times 12 \times 11 \times 10 \times 9 \times 8 \times 7}{1 \times 2 \times 3 \times 4 \times 5 \times 6 \times 7 \times 8 \times 9 \times 10 \times 11}$ = **12376**

13. 3 at $9.50
    2 at $5.50

14. Entering tunnel   =   3 seconds
    1 mile x 120 mph   = 30 seconds
                          33 seconds

15. 5 x 4 = 20 + 50% = 30
    ¼ x 20 = 5 + 50% = **7½**

16. **125**   100   164
              + 125   125
              225 = 15²   289 = 17²

17. **97 received $35 each.**
    97 and 35 are the only factors of 3395.

18. 40 hot dogs.

19. I was 16 and born in 1916.
    My grandfather was 66 and born in 1866.

CREATIVE PUZZLES: PUDDLED
(exercise on page 137)

Lay the stakes across the puddle, making sure they are an equal distance apart. Measure the length of each from top to bottom of the puddle, as indicated by the arrows, and add the measurements together. Divide the figure by the number of stakes lying across the puddle, and then multiply that figure by the width of the puddle to calculate the area.

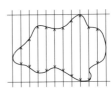

VERBAL EXERCISES
(exercises on page 138)

1. islander
2. **[c]** tiercel
3. **[c]** a whip
4. roundelay
5. pavid
6. sord
7. **[c]** sausage
8. canary
9. kennel
10. **[b]** marcasite

TECHNICAL EXERCISES
(exercises on page 140)

1. Mercury
2. **Factorial 6**
   6 x 5 x 4 x 3 x 2 x 1
3. Tangent
4. **6 out of 36 throws.**
   $\frac{6+4 \quad 4+6 \quad 6+5}{5+6 \quad 6+6 \quad 5+5} = \frac{6}{36}$
5. The formula for the surface of a ball is $\frac{4 \pi r^3}{3}$
6. Square base with sides half the height of the base.
7. Catenary
8. 8¾"   x 4¼"   x 2¾"
   220mm x 110mm x 70mm. 9¹ x 6¹ = **288 bricks** needed 9" x 3"
9. **6.** All snowflakes are based on an hexagonal shape.

TECHNICAL PUZZLES
(exercises on page 142)

1. **26 blocks**—only the middle block did not get painted.
2. 1110111.
3. lower than or equal to 4
4. r = radius,
   h = height, π = Pi
   πr2h
5. **37.** If there are 18 outside edges of a hexagon, the total must be 37.
6. copper and zinc
7. **[a]** icosahedron
8. stationary front

Looking Ahead

By now you have given your brain a thorough workout and have an insight as to whether you are a predominantly right- or left-brain person. You have also had the opportunity of going some way toward redressing the balance.

This will be something that you can build on in the future, as it is not possible to develop a balanced brain overnight just from a series of exercises. If after reading this book you are convinced that having a balanced brain is beneficial to you, then you will need to work out your brain regularly, in order to develop an even greater balance and not to let your dominant side take over again once you relax.

Your brain is undoubtedly your most valuable asset, but it is perhaps the part of our body that we most take for granted and is, therefore, most neglected. We need to exercise and care for our brain in the same way that we train other parts of our body. Just as gymnasts improve their performance and increase their chance of success at whatever level they are competing, you have at your disposal the mental gymnastics to give you the opportunity to increase the performance, and balance the laterality, of your own brain.

CREATE YOUR OWN WORKOUT

There are several ways you can do this. Some of the exercises in this book are easily adapted, such as thinking up different uses for everyday objects and creating drawings from geometric lines. To some, this may sound trivial, but anything that enables you to activate your brain cells and make you think can increase your brainpower.

There are other practical exercises you can carry out. Write out a series of instructions for performing an everyday task like getting your car out of your driveway in the morning, or making breakfast. You'll be surprised how difficult this is and how much you need to write down to record every single detail of the process. Again, it may seem simplistic, but you are getting your mind to work in a different or novel way, in the same way that taking a different form of physical exercise stretches muscles you did not even know existed.

←**IF YOU ARE DOMINATED BY THE RIGHT HEMISPHERE,** discipline yourself by making lists of tasks with deadlines.

During the past half century we have obtained a greater understanding of the workings of the human brain, and we should be able to use this newfound knowledge to our advantage, and to help society in general.

A great deal of the knowledge relating to brain hemispheres can be a considerable benefit in education systems. Until quite recently, left-handed children were encouraged to write with their right hands because it was thought that being left-handed would be a disadvantage throughout life. Unfortunately, this could only have the effect of repressing creativity in these children.

↑ **IT IS VITAL** to recognize and encourage creative talent in children from an early age.

←**THESE DAYS,** it is generally recognized that forcing left-handed children to use their right hands can stifle their creativity.

Looking Ahead

A newborn child instinctively uses both left and right hemispheres of its brain, and is able to learn an enormous amount of information and skills, including speech and language, without the use of books. However, this rapid development can slow down when the child enters the education system because it generally concentrates on the left side of the brain. So by the time the child reaches its teens, the right hemisphere has been taken over by the more dominant left hemisphere simply ecause it did not get enough opportunity to function.

↑ **LEONARDO DA VINCI**
The original Renaissance man:
creative genius, painter,
sculptor, architect, inventor,
and engineer.

On the other hand, our society is made of many specialists, and this is what our present education system is geared toward. It is a fact of life that specialists—especially if they become experts or leaders in their fields—can build up extremely successful careers and can provide comfortable lifestyles for themselves and their families. One problem with having a world of specialists is that each of us knows too little of the world in general—each knows a lot about one tree, but very little about the forest as a whole.

←INQUISITIVE, unpressured, and uncluttered, the mind of a young child is a mind at its most creative.

There are, however, signs that we are once again beginning to take a more holistic approach to our own lives, and to our planet in general, and there is now a swing back to the Renaissance man as typified by the fifteenth-century scientific genius Leonardo da Vinci.

To make the most of our precious assets, and to boost our own potential in everyday life, bear in mind the following seven *da Vincian Principles*—principles from which every one of us, whether we are right- or left-brain dominated, could benefit.

1. curiosity
2. willingness to learn from our mistakes and experiences
3. a continual refinement of the senses
4. embracing ambiguity and paradox
5. developing whole-brain thinking
6. cultivating grace, ambidexterity, fitness, and poise
7. discovering how all these principles fit together, and how everything connects to everything else

Index / Credits / Bibliography

CREDITS

Quarto would like to thank and acknowledge the following for supplying pictures reproduced in this book:
Key: b = bottom, t = top, c = center, l = left, r = right.
Imagebank p17t, p27 & p66; 66; Pictor International Ltd p8b, p13, p16/17c, p157t; The Art Archive/Tate Gallery London/Eileen Tweedy p10t; The Nobel Foundation p6t.
All other photographs and illustrations are the copyright of Quarto. While every effort has been made to credit contributors, we apologize in advance if there have been any omissions or errors.

BIBLIOGRAPHY

Greenfield, Susan *Brain Story* DK Publishing (2001)/BBC Consumer Publishing (2000)
Hancock, Jonathan *Maximize Your Memory* Reader's Digest (2000)/David & Charles (2000)
Ornstein, Robert *The Right Mind: Making Sense of the Hemispheres* Harvest Books (1998)
Reber, Arthur S. *The Penguin Dictionary of Psychology* Penguin USA (1996)/Penguin Books (1995)